Digital

Dedicated to:
Erna, my wife, and Bas, my son.

Colophon

ISBN:	978-90-5972-216-3 (1st Paperback Edition: 978-90-5972-203-3)
Publisher:	Eburon Academic Publishers, www.eburon.nl
Distribution:	University of Chicago Press, www.press.uchicago.edu/
Printer:	Lecturis, Eindhoven, www.lecturis.nl
Cover design:	Maarten van Schaik
Thanks to:	Excel Translations, Monique Kersten / Electric Words, Bob Emmerson
Copyright:	Jacob van Kokswijk © 2007
	Law School, University of Strathclyde, Glasgow, UK
	School of Culture Technology, KAIST, Daejeon-Seoul, KR
	Faculty of Behavioral Sciences, University of Twente, Enschede, NL

2nd Rev. Edition   V2.0 / 01-01-2008

Disclaimer All citations have been referred in all conscience and good faith. The URLs (hyperlinks) cited in this publication were active by access on 01-07-2007. Due to the transitory nature of some Websites, some of these links may no longer be active. Opinions expressed on these Websites are not necessarily those of the author. The author has already done as much as possible to retrieve the claimants on picture material. Nevertheless if someone should think they are entitled to an image, please get in touch with the publisher.

Copyright 2007: The copyright of the book belongs to the author under the terms of the United Kingdom Copyright Acts and Dutch Author Law.
Creative Commons: You are free to copy, distribute, display, and perform the work as long as you give the original author full credit. For any reuse or distribution, you must make clear to others the licence terms of this work. Any of these conditions can be waived if you get permission from the copyright holder. Nothing in this license impairs or restricts the author's moral rights.
Due acknowledgement must always be made of the use of any of the material contained in or derived from this publication. To discuss any aspect of the work you can contact the author directly, see WWW: <www.kokswijk.com>.

Digital Ego

Social and Legal Aspects of Virtual Identity

Jacob van Kokswijk

Eburon Delft
2007

Contents

	Author's Preface	9
	Foreword by Ian Pierson	11
	Abstract; Brief View to the Chapters	13
0	**Introduction**	**15**
1	**Identity and Identification in History**	**17**
	Introduction	17
	Identity in History	17
	From 'Neck Name' to Nick Fame	17
	I'm Not the Same	17
	Neck Name as Hyperlink	18
	Identity Politics	19
	Nickname and Fame	19
	Diminutive Nicknames	20
	Identity and Body	21
	Varying Identities	22
	Identity Concept	22
	Role Model	23
	Identity Standard	23
	Identity Change	24
	Identity and Imagination	24
	Identity and Inter Reality	24
	Plato's Utopia	25
	Interreality and Illusion	27
	Interreality Comparisons	27
	Pseudo Identity	28
	Boosting the Anonymity	28
	Masked Identity	28
	Switching Identities	29
	Illusory Worlds	29
	The Value of Anonymity	30
	Sick or Side?	31
	Identity and Identification	31
	Identity Construction	31
	Identification by Identity Control	31
	The History of the Passport	32
	Biometric Identity Control	33
	Summary Chapter 1: 'Identity and Identification in History'	34
2	**The Impact of Technology**	**35**
	Introduction	35
	Empowering the Imagination	35
	Reality and Truth	35
	Imagination 'Powered by Technology'	36
	Fantasy Empowered	36
	From Connection to Common Use	37
	Tele-technology	37
	Shocking Experiments	37
	Télégraphe	37
	The Victorian Internet	37
	Wireless	38
	Invisible Visible	39
	Cybernetics	39
	Cyber-Reality	40
	Cyberspace	41
	The Next Ten Years	42
	Technical Background of Virtual Identity	43
	Identity in Information Technology	43
	Online ID	43
	Network Identification	44
	Connecting the Identity	45
	Enjoy the Anonymity	46
	To a Safe Domain	46
	Online Identification	47
	Online Identity and Privacy	49
	Online Secured Access	49
	Identity Management Infrastructures	49
	Web Address as Human ID	50
	Summary Chapter 2: 'The Impact of Technology'	50

3	**Virtual Identity and the Internet**	51
	Introduction	51
	From Pseudo Plato to Baby.com	51
	The Individual	51
	Online Identity	51
	Virtual Entities	52
	Cyber-Entities	52
	Virtual Identity	53
	Everyone an Identity	53
	But Why Virtual Identities?	53
	Who Am I?	54
	Makeability	54
	Online (Mask) Identity	55
	The Origin of a Virtual Identity	56
	The Shape of a Virtual Identity	56
	Virtual Life	57
	Theoretic Background of Virtual Identity	58
	Disembodied Information	59
	Social Psychological Background of Virtual Identity	60
	Ethical Background of Virtual Identity	60
	Integration and Separation of Identities	61
	Offline and Online Experiments	61
	Identities in All Kind of Varieties	61
	Integrating Offline / Online	62
	Identity Construct	63
	Summary Chapter 3: 'Virtual Identity and the Internet'	64
4	**Social Aspects of Virtual Identity**	65
	Introduction	65
	Signals from Society and Media	65
	Media Technology as an Extension of Man	65
	The Rate of Technology in the Changing Society	65
	Emotional Machine	66
	Value and Importance of Virtual Identity for Society	67
	The Organism is the Message	68
	Social Paradigm Change	68
	Everybody Can Test You're a Male Dog!	69
	Fantasy, Media, and Identity	70
	The Media Meet the Need for Truth	70
	The Media Meet a Need for Experience	71
	Fantasising Meets Human Needs	72
	Categorisation of Fantasy Needs	73
	Real Imagination in Virtual Affairs	74
	Business as Booster	74
	Identity, Time and Origin	74
	Social View to Internet	74
	Time in Virtual Space	74
	Time is a Factor	75
	Being Up 2 Date	75
	Time has Many Faces	76
	Time and Identity	77
	Time is the Message	77
	Emerging Technologies	78
	Summary Chapter 4 'Social Aspects of Virtual Identity'	78
5	**Forms, Varieties and Valuables of Virtual Identities**	79
	Introduction	79
	Environment of Virtual Identities	79
	Intelligent Agents	80
	Use and Users of Virtual Identity	80
	Use of Virtual Identity	80
	Online Conversation Modelling	80
	Online Chat	81
	Online Intelligent Agent	82
	Online Search Agent	83
	Online Instant Messaging (IM) Agent	83
	WebBot – Virtual Assistant on the Web	84
	Botnet(work)	84
	Virtual Assistants of the Web Criminals	84
	Offline Games	85
	Online Games	85

	Online Presentation	86
	Online Communities	86
	Online Dating	87
	Online Interactive Broadcasting	87
	Online Interactive Storytelling	87
	Online Worlds	88
	Online Machinema	88
	Online Virtual World Movie	88
	Online Relational Networks	88
	Online Self-service	89
	Online Feelings and Satisfaction	89
	Online Intimacy	90
	Online Body	90
	Online Death	91
	Online Resurrection	92
	Online Transactions and e-Commerce	92
	Online Property	92
	Online Marketplace	92
	Economic Emerge of New Services	93
	Online Tax	93
Users of Virtual Identity		93
	Human	93
	Machine	93
Avatar Rights		94
	Effects of the Use of Virtual Identity	95
Summary Chapter 5 'Forms, Varieties and Valuables of VIDs		95

6 Virtual Identity and Anonymity — **97**

	Introduction	97
Anonymity and Pseudonymity		97
	Anonymity, Technology and Virtual Identity	97
	Anonymous Identity in the Information Age	98
	Online Anonymity	100
	Personal Profile in both the Environments	101
To Be or Not to Be Identified		101
	Political View to Identity @ The Internet	101
	Privacy and Identity	101
	The Next Mobile Revolution	103
	New Rules about Data Mining	103
	From Anonymity to Pseudonymity	105
	Identification, Authentication and Data	106
	Laws of Identity	107
	De-perimeterisation	108
Summary Chapter 6: 'Virtual Identity and Anonymity'		108

7 Legal Issues and Policy of Virtual Identity Part I — **109**

	Introduction	109
The Crawl of Codes		109
	The Progress of Order and Law	109
	Thou Shalt Not Make for You an Image	109
	According to what law are you living?	111
Sovereignty and Jurisdiction		112
	Borderless Law and Jurisdiction	112
	Cross-border Law and Jurisdiction	114
	Wrapping-up the Cases	117
The Internet and Sovereignty		117
	Law Will Regulate All?	118
	Cyberspace is Not Outer Space	118
Can/Should the Internet be Regulated for Identities?		119
	The Need and Distaste for Rules	119
New Media, Renewed Law		119
	Licence to Chill	119
	Control without Government	121
Control by Code		122
	The Palette of Codes	122
	Code to Rule the Citizen	122
	Code is Control	123
	Code and Commerce	124
	Law as Country Code	124
Summary Chapter 7: 'Legal Issues and Policy of VIDs'		126

8 Legal Issues and Policy of Virtual Identity Part II — **127**

	Introduction	127
	Real Behaviour in Virtual Environments	127
	Behavioural Control in Virtual Society	127
	Social Control	127
	Status of Virtual Identity in Law of Persons and Property	129
	Hidden Helpers	129
	VID in Law of Persons	129
	Humanitarian Law	131
	VID in Law of Property	132
	The Position of the Virtual in the System of Rules	134
	Interaction and Reciprocity	134
	Customary Law in Cyberspace	135
	Economy and Law	135
	Hidden Habitants in the Heat of Hyperlinks	136
	The Attraction of Anonymity	136
	Can an Anonymous VID be Anonymous	137
	The Legally Intangible Anonymous ID	138
	Traces of Terminal Use	139
	Use of VID as a Cover for Criminal Conduct	139
	Follow the Virtual Money to the Crime	139
	Summary Chapter 8 'Legal Issues and Policy of VID II'	140
9	**Legal Consequences of Virtual Identity**	**141**
	Introduction	141
	Footboard to Virtual Footprints with Real Foothold	141
	Artificial Agents	141
	A Waver of Autonomic Executing Agents with VID	141
	Identity Generator	141
	Who Did It ?	142
	Who Did What?	144
	Legal Difference between an Identity in Cyberworld	144
	Marking if Difference and Exclusion	144
	Virtual Identity in the Legal System	146
	According to the Conditions of Our World	146
	Code, Code or Code	146
	The Range of the Law	147
	The Virtual Party in the Circuit	148
	My Thinking VID	148
	My Stolen VID	148
	My Naked VID	149
	My Suspected VID	150
	Online Penalties	151
	Cybercrime, What's Next?	151
	Limits to Regulation	151
	The Code and the Coders	151
	The Coders	152
	Summary Chapter 9: 'Legal Consequences of Virtual Identity'	152
10	**The Way Forward**	**153**
	Introduction	153
	The Code that Rocks the Cradle Rules the World	153
	New Legal Definitions for New VID Concepts	153
	Virtual Law	153
	Customary Law and Arbitrage	154
	Future Expansion and Expectations	155
	Legal View	155
	Golden Code	156
	Conclusions	156
	1) Jurisdiction	156
	2) Identity Correlation	156
	3) Identity Switch	156
	4) Autonomous VID	156
	5) Freedom of Thought	157
	6) Law Making	157
	7) Public Passkey	157
	Summarized	157
	Recommendations (With Attention to the Legal Topics)	157
	Summarized	158
	Final	158
11	**Appendices**	**159**
	Bibliography of Related URL's	159

Schizophrenia as Commodity Fetish

'A person called Julie was presented on a computer conference in New York in 1985. Julie was a totally disabled older woman, but she could push the keys of a computer with her headstick. The personality she projected into the "net"--the vast electronic web that links computers all over the world--was huge. On the net, Julie's disability was invisible and irrelevant. Her standard greeting was a big, expansive "HI!!!!!!" Her heart was as big as her greeting, and in the intimate electronic companionships that can develop during on-line conferencing between people who may never physically meet, Julie's women friends shared their deepest troubles, and she offered them advice--advice that changed their lives. Trapped inside her ruined body, Julie herself was sharp and perceptive, thoughtful and caring. After several years, something happened that shook the conference to the core. "Julie" did not exist. "She" was, it turned out, a middle-aged male psychiatrist. Logging onto the conference for the first time, this man had accidentally begun a discussion with a woman who mistook him for another woman. "I was stunned," he said later, "at the conversational mode. I hadn't known that women talked among themselves that way. There was so much more vulnerability, so much more depth and complexity. Men's conversations on the nets were much more guarded and superficial, even among intimates. It was fascinating, and I wanted more." He had spent weeks developing the right persona. A totally disabled, single older woman was perfect. He felt that such a person wouldn't be expected to have a social life. Consequently her existence only as a net persona would seem natural. It worked for years, until one of Julie's devoted admirers, bent on finally meeting her in person, tracked her down.' [1]

---

[1] Source: AR Stone, *Will the Real Body Please Stand Up?* in M Benedikt (ed), 'Cyberspace: First Steps' (MIT Press, London 1994) p83 (81-118); WWW: < http://molodiez.org/net/real_body2.html>. See note 166, page 51.

Author's Preface

'Telecommunication divides more people than communication connects.' [2]

Most of you will be familiar with the alias of your email account. Pseudo identities (pseudonyms) are used more and more in electronic systems and networks. Your log-in username for your email or your nickname for participating in an online community is a kind of virtual identity too.

During my research (2000-2003) I became familiar with the so-called virtual identities in software, used for – mostly temporary – unique actions during the executing process, e.g. as "intelligent agents". In my latest research into the human behaviour in cyberworld I was amazed by the amount of so-called "nicknames", a pseudo(nym) that is created when you enter the chat rooms, virtual societies and online games environments. I discovered that people were using different pseudonyms at the same time, depending on the situation and mood. Some of them also change their gender in these virtual worlds, often without being recognised. (see cited phrase p 8) Once eBay and hundreds of other e-markets started their online marketplaces, millions created a pseudonymous trade name for online business and the phenomenon of self chosen virtual identities was soon widely spread.

In the past only artists, criminals and some peers created and used pseudonyms for their actions. Today, children and adults can be Superman in real action and performance. Now that space and time have fallen away, all traditional connections and identities are suppressed on the Internet, both the individual and the collective ones. New forms of anonymity, gender and identity switches are cultivated on the World Wide Web, thus superseding the idea of a global village. Watching the development of the mixed reality – both the virtual and physical worlds are seamlessly mixed into a hybrid human environment called "interreality" – we can be sure that the current ideas and rules about virtual identity will have to be changed, step by step. New forms of anonymity and dissociation (assuming multiple personalities) are cultivated in this virtual society, leading to different behaviour patterns. Much attention is paid to this phenomenon from a sociological viewpoint in the academic world but less so from other single and multidisciplinary research programmes.

The aim of this book annex research report is an exploration, a quest into developments in society that are decisive for integrating new communication technologies, into questions and suppositions that have arisen along the way, as well as into research that has been conducted worldwide in the past, leading to recommendations. These issues, the relationship between identity and technology, society, law and innovation is further worked out. In my explorations I have looked at the phenomenon of 'the virtual identity' from different public sides, taking into account the various scientific perspectives, such as social psychology and philosophy, as well as the legal and technological knowledge areas, with some emphasis on the electro(nic) technical (information and telecommunication) and social psychological sciences.

Glasgow, July 1st, 2007
Rev. January 1st, 2008

Jacob van Kokswijk

---

[2] Jacob van Kokswijk, PhD, MSc, MO; Expert of human behaviour in digital media; WWW: <http://www.kokswijk.com>

'The wisdom of the crowd can quickly become the stupidity of the mob.'

Kevin Kelly, founder of Wired magazine

Foreword by Ian Pierson

'Anyone can predict stuff, but only a few get it right' [3]

In the mid 1990s, I used to spend several hours a day in chat rooms. I had several different characters ('identities') that I used, basically just acting out different roles for fun. The more serious side of such identities is that other people don't always know you are just pretending and relationships can form with people who don't exist. What starts out as a frivolous and harmless recreational activity can sometimes have a far-reaching impact. So I often wondered about the impact of our virtual identities on the rest of the world. Of course the online environments have since then evolved very significantly.

So now, thankfully, Jacob van Kokswijk has produced this book, looking at the whole field in detail and pondering the legal consequences. It is very timely. The virtual world has taken off as a socialisation environment, with millions of people interacting in virtual worlds. It is even becoming a business environment. Far more is at stake than just casual chit-chat and minor misunderstandings. As artificial intelligence progresses, our virtual identities will come to include agents that act like us and for us even when we aren't present. The avatar will have intelligence behind it.

But as the power behind the agents becomes more powerful, so does the risk. What happens when someone steals our identity, or simply corrupts it, making it behave slightly differently than we intended, for their own purposes? How much can we control it in an age with ever more sophisticated hackers and criminals? If our agent says something it shouldn't, who is responsible? And if our agent is constantly changing, or popping in and out of existence under different guises, what then? There are many legal questions that pop up.

Jacob explores these and many other questions and demonstrates very clearly just how important a field this will become, taking us from today's crude virtual environments to the more deeper issues that will arise as they become more interwoven into our lives. And he does so with calm, arguing well that we should not panic and introduce poorly thought through reactive legislation, but to think carefully before we act.

Because the future will not be as simple as today. AI agents will not always have a human directly responsible for them. Many will evolve and emerge over time through complex and highly dispersed interactive processes, and yet will inhabit the same virtual worlds as people and their avatars. These worlds will not exist neatly on a single machine but be distributed across many countries. Their virtual populations will span the whole globe. But Jacob empathises that such changes of the technology platforms and capabilities are really just mere distractions, and that it is far better to create sets of lasting social principles than detailed legislation, and thereby to allow emergent communities to govern themselves by peer pressure and a sense of right and wrong rather than blind rules that will date quickly. By introducing calm and clear thinking where many are panicking, Jacob's book makes a very valuable contribution.

IAN PEARSON

---

[3] Ian Pearson, Futurologist BT Group Chief Technology Office; WWW: <http://www.btinternet.com/~ian.pearson/>.

'My name is Bond... James Bond.'

Sean Connery (1930-) Scottish actor

Abstract

All this varied text contains a connecting thread: **virtual identity**. *A Virtual Identity is the representation of an identity in a virtual environment. It can exist independently from human control and can (inter)act autonomously in an electronic system as well as in a global network. Virtual identities are not a novelty; nor are words such as cyberian and interreality neologisms. Almost 2.500 years ago people used pseudonyms to be anonymous, and imaginary worlds were discus-sed as well as the jurisdiction of law. For more than 150 years there was a de facto Internet, with chat rooms, virtual affairs and online identities. Going back in history, discussions were also held about the shameless youth that used technology, about rude colonists who made their own rules in the conquered territories, and about floating people who were hopping between imaginary and realistic societies. A virtual identity is the same commodity as money.*

Millions of people around the world duplicate their identity in order to inhabit virtual worlds: cyber societies and multiplayer online games where characters live, love, discuss, trade, mask, cheat, steal, and have every possible kind of adventure and transaction. Far more complicated and sophisticated than early the discussion groups and video games, people can now spend countless hours in virtual universes like SecondLife, creating new identities, falling in love, building cities, making rules, and breaking them. Using online identities manually in cyberworld result in a range of virtual identities. Even when there is no direct user (human) controlled action of your 'second ego' (initiated by yourself) the computers and connected networks will be busy with executing your orders in the virtual world.

Virtual IDentities (VIDs) can also live their own life. The use of autonomous virtual identities such as your chess robotic opponent in a game of chess is increasing. More and more software contains parts of artificial intelligence that uses the VIDs. The new service oriented applications will deliver and present the user (customer) with a package of software, code and content, assembled and personalised by artificial intelligence (AI) 'powered' software agents. One could say that there is always a real person somewhere as manager and operator behind the screens, the practical situation is such that the software's far-reaching autonomy is designed to run automatically and only to make an occasional choice in content management, individual needs (including the presentation requirements for the used devices) and the service offer, in order to achieve the quality of service that is agreed between the user and the provider (manufacturer). The AI-generated, modified, – sometimes manipulated – and removed ad hoc 'intelligent' (software-)agents behave themselves during their life as virtual identities. This phenomenon raises questions about the legal consequences. If non-human virtual identities can live their life as such virtual aliens, who is responsible and liable for their actions in networks and systems on our earth? Every investigator will question in advance 'who did what where when and how?' Many politician will ponder on 'what shall we do with the virtual citizens?'.

In this book the different forms of identity are identified, including the various ways identities are used and practised in the computer environment and in today's society. One cannot prevent theft. You cannot fight windmills. Criminality is part of our society, and we have to tolerate and live with that. We are not able to prevent and solve every criminal attempt in a theoretical or technical way. Every person runs the risk in our society. Not that we don't have to do something against criminality but the policy should be realistic, not a frenetic searching for impossible solutions. Since the 1980s the constant worldwide electronic monitoring has emerged. It may prevent crime but at what cost? The desirability is seldom considered with the high impact on privacy and the poor benefits. As Popper stated in his 'Objective Knowledge, an evolutionary approach': 'A theory is not stronger by more technical evidence but by the more that theory can resist the attempts to refute it'.

So, we have to consider that – even when we shall have cyber cops and virtual investigators – we cannot regulate all, as their equivalents in the real world neither could. It also seems that each new media is more regulated than its predecessor(s). With the new media technologies, the choice of law issues (such as domicile) may become crucial as the exploitation of identity can easily and quickly occur at national levels in ambiguous jurisdictions. A well-communicated consensus on the 'virtual' law would promote certainty and would reduce court (shopping) cases.
As in history new technology is regulated by old law. Equally, most elements of the Internet usage and service content were regulated in some form or fashion – prior to the arrival of the Internet.

Despite the calls by some for the development of Internet-specific law, or cyberspace law – similar to the Law of the Sea – the information technology is changing so rapidly that for any 'sui' generic body of law that is developing, implementing and maintaining, it is a lost race against the 24/7 Internet time. Despite the technology pushed idea that the Internet culture is the same for all contributors, there are and will be local and cultural differentiations in compliance with the law. The 'ad hoc' way of gradual adaptations of the tried and tested fundamental legal principles, as we have seen in free speech, economics, and in privacy and intellectual property protection, is likely to be more successful.

The ways are examined in which laws can be used to create positive ethical models in individuals and groups. It is well known by policymakers and lawmakers that by passing a law the form of societal control which restricts the undesirable behaviour is very important. If the law becomes more widely accepted, people begin to reduce misbehaviour on the principle that it is `wrong' to do so. However, the makers of policies and laws are seldom aware of the societal structure of 'cyberspace', and for this reason there is the danger that the laws they make will not create the desired ethical model, but will instead create a backlash or revolutionary movement against the society. By observing the human behaviour in virtual communities and by continuing to take time to develop realistic policies, legal ways of mediation, and effective laws, it is possible to avoid such a backlash.

Summarized in the conclusion of this survey is that, just as in the real world, a virtual gold rush of virtual identities will set the exploitation codes for data mining and transactions. Disputes will be solved by a mix of codes, submitted by the experts, the peer groups and the environmental conditions. Codes should be public and free to be empowered and improved by the community. The issue of virtual identity, including the rights of avatars and intelligent agents, constitute legal challenges for our online future. Though virtual worlds may have started as games, they are rapidly becoming as significant as real-world places where people interact, shop, sell, and work. As society and laws begin to develop within virtual worlds, we need to have a better understanding of the interaction of the laws of the virtual worlds with the laws of this world.

The recommendation is that just as in the real world the expertise of the citizens in virtual worlds can be used to let the people substantiate and support the law and sentences. Lawmakers and courts also have to transform themselves to virtual imagination. Complex illegal looking acts by autonomous virtual identities could be handled as effects of natural imperfection, even when it is self-starting after human intervention. Finally, there is no code without coders. They have to be on one code.

Brief View to the Chapters

Chapter
1. is about the history of identity and identification, politics and imagination. It names masking, constructing, changing, varying and playing with identities.
2. explains the relation with technology that empowers the imagination and enables the management of the continuous changing identity.
3. shows how the Internet emerges with the individual identity. It explains the integration and separation of someone's identity in systems and how this leads to disembodying the identity.
4. points the social aspects of Virtual IDentity (VID) and the calls the established use of VIDs in our society.
5. explores the forms, varieties and valuables of the and lists the broad use and users of the VID.
6. focuses on anonymity and pseudonymity in the real society. It explains the catch for surveillance and points out the topics that are important for economics.
7. concerns the legal issues, such as the myriad of codes, sovereignty, and regulations for identity. It highlights the electronic distributed code to control the owner of an identity.
8. describes the position of the virtual identity in both the virtual and real world's system, and the tangibility of an anonymous ID when used improperly or legally.
9. focuses on the legal consequences, such as the differences between a real and a virtual ID, the parties involved, the regulation by codes, and names the online penalties by offence.
10. leads to the way forward, with new VID concepts and new legal perspectives, and finishes with some conclusions and recommendations.

0 Introduction

'There are no passengers on spaceship earth. We are all crew.'

Marshall McLuhan [4]

This book discusses the changes in information and communication thanks to tele-technology and is also a quest to schedule the new developing human behaviour and social relations acting with the Internet and mobile technology on the legal agenda.

The scope of this book focuses specially on to the phenomenon of **'virtual identity'**, a machine-related creation of an anonymous reproducible identity that exists mainly in virtual worlds in the shape of an 'agent', and has – for example – relationships with the online identity of a human being. The research is internationally focused and points out – more abstractly – the relationship between existing law and new technology. It relates to the privacy paradox in the private versus public space (such as web 2.0) and to social networking privacy issues.
Necessarily a broad multidisciplinary research has been conducted into which perspectives were required and desirable with regard to this 'virtual identity', but in order to keep to the broad outlines further in-depth differentiation was not possible.

> *'It is puzzling that even theorists steeped in this tradition seem slow to see the relevance of new information technologies to their preferred ways of understanding development. Shifts in the way that children are relating to knowledge may be happening faster than we are organizing a research agenda.'*
>
> Charles Crook [5]

The Internet – since 1991 an open worldwide network – is pioneering 'do it yourself' processes. The youngest generations were quick to latch on. Why wait for the deliberations of the traditional world, which avoids risks but keeps the money earned itself when successful? Why do you always have to be recognised as an adolescent when you are doing adult actions? With your mobile phone, a borrowed HTML [6] script and a few hours behind the PC in the library you position yourself in the world. You are an artist, a photographer, architect, publisher, media station and movie director for and by yourself at the same time. The virtual world seems a mirage to some, but it is not. Also, the virtual identity seems to be an autonomous identity that can but does not have to be related uniquely nor singularly to a human being. One aim of this research report is to draw attention to, and to stir up a discussion in scientific circles about this phenomenon of virtual identity that is frequently used in the Internet environment.

Apparently we find ourselves at the edge of science with a theme into which broader and more in-depth research should be carried out in order to gain more insight into the effects of communication technology, technology that has been changing at full speed over the past few decades. The fast developing integration of so called 'intelligent agents' (popular in e-commerce applications) also requests a thorough examination of all legal aspects.

The core knowledge related to this technology is spread among specific interscientific communities, such as Computer Science, Telecommunications, Human-Computer Interaction, Computer Supported Cooperative Work and Artificial Intelligence. And furthermore among cognitive studies, communication studies, ergonomics, industrial psychology, group dynamics, anthropology, philosophy, social psychology, organisational psychology, sociology, and the law.[7] Optimum use of this knowledge is hampered by the considerable fragmentation between the various sciences which – in my experience – do not have much contact with one another. But they all have a connecting thread: **virtual identity**.

---

[4] M McLuhan (1911-1980). WWW. <http://www.marshallmcluhan.com/>.

[5] Charles Crook, 'Cognitive development and Internet' (*Monitor on Psychology*, Loughborough Univ, 2000) Vol. 31/4.

[6] HTML, short for HyperText Markup Language, is the predominant markup language for the creation of web pages. It provides a means to describe the structure of text-based information in a document - by denoting certain text as headings, paragraphs, lists, etc. - and to supplement that text with interactive forms, embedded images, and other objects.

[7] Law topics such as copyright, identity, privacy and digital government: in this case private law, public law and international law, including tax law, information law and the right to vote.

Virtual identities are not a novelty, neither are words such as "cyberian" or "Interreality" neologisms. Almost 2.500 years ago people used pseudonyms to be anonymous, and imaginary worlds were discussed as well as the jurisdiction of law.
For more than 150 years there was a de facto Internet, with chat rooms, virtual affairs and online identities. Going back in history, discussions were also held about the shameless youth that used technology, about rude colonists who made their own rules in the conquered territories, and about floating people who were hopping between imaginary and realistic societies.

Cyberworld seems to expose the same colonialism as in the past (e.g. America), with three differences:
1) you have to (and you can) develop your "new world" from scratch (digital space is empty; there is no ground like in the real world; you have to design something like "ground" if you need to: for instance, in virtual reality a building can float);
2) you are with your mind in cyberworld but with your body in a physical world (however, senses of your body are connected);
3) cyberspace doesn't have (as far as we know now) its own materials or alternative energy sources in order to maintain itself and survive.

1 Identity and Identification in History

> '*A musician must make music, an artist must paint, a poet must write, if he is to be ultimately at peace with himself. What a man can be, he must be.*'
>
> Abraham H. Maslow [8]

Introduction

From the Inca's [9] in the Andes to the Zhou [10] in China, and from ancient Greece to modern Great Britain, the continual changes in personal identity of people lead to forms of identification, including marks, tattoos and artefacts. Some people split their personality and were judged to be sacred or bewitched. As modern science came to the fore, the primary concerns were how to identify the real identity. Today, a person has to present a chip card or identify themselves biometrically to 'prove' that they are the legitimate user of the card.

This chapter is separated into five subs: Identity in History, Varying Identities, Identity and Imagination, Pseudo Identity and Identity and Identification.

Identity in History

The particular subtitle points out the History in 'Neck' and Nick names, and Identity Politics. Nick names and their relation to both body and identity will be explained.

From 'Neck Name' to Nick Fame

I'm Not the Same

Greek theatre and plays have had a lasting impact on Western drama and culture.[11] In the 500s BC, a poet named Thespis[12] is credited with innovating a new style in which a solo actor performed the speeches of the characters in the narrative (using masks to distinguish between the different characters). The actor spoke and acted as if he was the character, he also interacted with the chorus, who acted as narrators and commentators. Thespis is therefore considered the first Greek "actor," and his style of drama became known as tragedy.

In ancient Greece, one of the earliest indications of interest in the problem of personal identity occurs in a scene from a play (written in the fifth century BC by the comic playwright Epicharmus [13]). It takes the form of an exchange between two characters, one of whom has no funds but owes the other money.[14] The one without funds – the debtor – claims that just as a pile of pebbles from

---

[8] Abraham Maslow (1908-1970) is the psychologist praised for his work on human needs. He illustrated our hierarchy of needs.

[9] The Inca (or Inka) Empire was the largest empire in pre-Columbian America, and at its peak had a large empire by world standards. The administrative, political and military centre was located in Cuzco. The Inca Empire arose from the highlands of Peru sometime in early 13th century. From 1438 to 1533 AD, the Incas used a variety of methods, from conquest to peaceful assimilation, to incorporate a large portion of western South America, centred on the Andean mountain ranges. WWW: <http://en.wikipedia.org/wiki/Inca>.

[10] The Zhou Dynasty (Chinese: 周朝; 1122 BC to 256 BC) followed the Shang (Yin) Dynasty and preceded the Qin Dynasty in China. The Zhou dynasty lasted longer than any other in Chinese history; though the actual control of China by the dynasty only lasted during the Western Zhou. The dynasty spans the period in which the written script evolved from the ancient stage as seen in early Western Zhou bronze inscriptions, to the beginnings of the modern stage, in the form of the archaic clerical script of the late Warring States period. During the Zhou, the use of the Book of Jiuxing criminal laws was introduced to China. WWW: <http://en.wikipedia.org/wiki/Zhou_Dynasty>.

[11] Greek Drama is a theatrical tradition that flourished in ancient Greece between 600 and 200 BC. The city state of Athens, the political and military power in Greece during this period, was the epicenter of ancient Greek theatre. Athenian tragedy, comedy, and satyr plays were some of the earliest theatrical forms to emerge in the world.

[12] Thespis of Icaria (6th century BC) is claimed to be the first person ever to appear on stage as an actor in a play. WWW: <http://en.wikipedia.org/wiki/Thespis>.

[13] Epicharmus [540 - 450 BC] was a Greek dramatist and philosopher often credited with being one of the first comic writers in the classical Athens. Socrates refers to him as 'the prince of Comedy'.

[14] Early Greek Comedy and Satyr Plays. Section 3, Chapter 8. Pointed out by: D Sedley, *The stoic criterion of identity* (Phronesis, 27 1982): 255-75, p. 255. WWW: <http//www.usu.edu/markdamen/ClasDram/chapters/081earlygkcom.htm>.

which one pebble has been removed it becomes a different pile of pebbles; anything that undergoes any sort of change thereby becomes a different thing. The one to whom the debtor owes the money – the lender – agrees. 'Well, then', says the debtor, 'aren't people constantly undergoing changes'? 'Yes', agrees the lender. 'So', the debtor concludes triumphantly, 'I owe you nothing since I'm not the same person as the one who owes you money'. Exasperated, the lender hits the debtor, who then protests loudly that he has been physically abused. 'Don't blame me', replies the lender, 'I'm not the same person as the one who hit you moments ago'.

Like many philosophical puzzles, this brainteaser concerning identity, responsibility and egoistic concern, is common for Greek intellectuals and artists to believe that everything is in a state of constant flux, including human beings. Some of them recognised that if all things are in constant flux, then arguably no one lasts for long and, hence, there is no metaphysical basis for responsibility and self-concern.

Martin and Barresi [15] argue that the interesting – borderline amazing – thing about this ancient scene is that it suggests that even in fifth century BC Greece, the puzzle of what it is about a thing that accounts for its persisting over time and through changes could be appreciated even by theatre audiences. Another interesting thing about the scene is its more specific content: both debtor and lender have a point. Everyone is always changing. So, in a very strict sense of 'same person', every time someone changes, even a little, he or she ceases to exist: the debtor is not the same person as the one who borrowed the money, the lender not the same person as the one who hit the debtor. This very strict sense of 'same person' is not an everyday notion but the product of a philosophical theory. It is also not a very useful sense of 'same person' – unless you owe someone money! According to Martin and Barresi, in everyday life we want to be able to say such things as, 'I saw you at the play last night,' and have what we say be true. If everyone is constantly changing and every change in a person results in his or her ceasing to exist, no such remarks could ever be true. Assuming that such remarks are sometimes true, there must be a sense of 'same person' according to which someone can remain the same person in spite of changing. Saying what this sense is, or what these senses are, is the philosophical problem of personal identity.

Neck Name as Hyperlink

The Incas (<1572 AD) used number values and relations to encode information on their familial, social and political relations and structures, as well as concerning type and quantity of property (such as cattle). It was, for example, used by the tax collector to record the type, the quantity and the place where tributes were collected. Their arithmetical practice had been based on a well developed basis of philosophical values and principles, which reflects to maintain a continuing effort for harmony and balance of the material, moral and social environment in the group. (Urton et al) [16] The system of accounting relied on the *quipu*. The Inca people wear the personal quipu around the neck as an identity proof. (Figure 1)

This type of number values, appreciations and coupled relations is now in use by hyper links and search engines on the Internet. Information is tied to a link with which – without knowing – the Internet surfing individual betrays himself. In the received string can be read by which way and with which interest the specific netter has terminated (if the recipient is in the possession of required equipment and information).

Figure 1: The Inca people wear the personal quipu around the neck as an identity proof.

Social Identity Registration

'*Cords of various colours were attached to a main cord with knots. The number and position of knots as well as the colour of each cord represented information about commercial goods and resources.*

---

[15] R Martin and J Barresi,
The Rise and Fall of Soul and Self, An Intellectual History of Personal Identity (Columbia University Press, New York 2006).

[16] G Urton and P Nina Llanos, *The social Life of Numbers. A Quechua Ontology of Numbers and Philosophy of Arithmetic* (University of Texas Press, Austin 1997).

Quipu means knot in Quechua, the native language of the Andes. The quipu was also useful for census taking and provided a mass of statistical information for the government. The Inca people wear the personal quipu around the neck as an identity proof. Knotted numbers are fundamental to the symbolic system of the quipus. Messengers could carry a quipu from Quito to Cuzco in three days, less time than it sometimes takes by car. Archaeologists are now suggesting that authors used the quipu to compose and preserve their epic poems and legends. Because there were relatively few words in Quechua, they could be used as pronunciation keys on the cords. Then each knot on a cord designated a syllable of the word represented at the head of the cord. For example, the name of Pachacamac, god of earth and time, was divided into four syllables. So if two knots were tied close to the key word, the author had written the word 'pacha' or 'earth.' But if the two knots were tied further down the cord, they indicated the last two syllables of the god's name and meant the word 'camac' or 'time'. We still don't know exactly how to use the quipu. The significance of the knots and colours remains a mystery.' [17]

Identity Politics

Culture and identity play a vital role in processes of social transformation.[18] Identity politics was (and still is) often a form of mobilisation against globalising forces which appear as threats to the livelihoods and values of marginalised groups.

Identity is bodyless. In almost all faiths the created personalities also die and resurrect, sometimes split or merge, depending on the circumstances. People like to play with life and death and identities in order to cope with the feelings of the great unknown.

Nickname and Fame

In Viking[19] societies, many people had nicknames '*heiti*', '*viðrnefni*' or '*uppnefi*' which were used in addition to, or instead of their family names. In some circumstances the giving of a nickname had a special status in Viking society in that it created a relationship between the name maker and the recipient of the nickname, to the extent that the creation of a nickname also often entailed a formal ceremony and an exchange of gifts. In Middle English the word "nickname" was "ekename" (from the verb to eke, "enlarge"; compare Swedish *öknamn*). [20]

Later, an ekename developed into a nickname when the "n" shifted through *junctural metanalysis*. In daily life a nickname is a short, clever, cute, derogatory, or otherwise substitute name for a person or thing's real name, e.g. "specs" for a person who wears glasses. Sometimes related to an 'alias', a personal nickname may be sarcastic, simply ironic, or a reference, e.g., "curly" for someone with straight hair (or no hair at all). This form was typical in Australian English in the mid 20th Century but less so in current parlance, e.g: "shorty" for a very tall person. A nickname is sometimes considered desirable, symbolising a form of acceptance, but can often be a form of ridicule. As a concept, it is distinct from both a pseudonym and a stage name, and also from a title (for example, City of Fountains), although there may be overlap in these concepts.[21]

Even 'popular' laws and jurisprudence became and still become nicknames. Already in 1914, the US Committee on Legal Bibliography reported:

> *'Many of the laws upon our statute books are known generally by a nick-name or have a popular title; e.g., Bland-Allison act; Hepburn act; Sherman law, etc. Your Committee would recommend that a resolution be prepared in the name of the Association and addressed to the proper official or department having the preparation of such laws in charge, urging the importance of the publication of a pamphlet giving references to each of the laws called or known by a popular title. Also, we ask that the indices to subsequent Statutes-at-large and Revised Statutes include citations to any and all laws passed by the Congress of the United States that may be known by a popular title.'* [22]

---

[17] M Ascher and R Ascher, *Mathematics of the Incas: Code of the Quipu* (Mineola, NY 1997) pp 13-35. Photo by Marcia and Robert Ascher in the Museo National in Lima, Peru; WWW: <http://instruct1.cit.cornell.edu/research/quipu-ascher/photos/lima/index.htm>.
'Quipu: A Modern Mystery'; WWW: <http://www.sfu.ca/archaeology/museum/laarch/inca/quipue.html>.

[18] M Castells, *The Power of Identity* (Oxford, Blackwells 1997) pp 65, 143.

[19] The term Viking commonly denotes the ship-borne warriors and traders of Norsemen (literally, men from the north) who originated in Scandinavia and raided the coasts of the British Isles and mainland Europe as far east as the Volga River in Russia from the late 8th-11th century. This period (793–1066 AD) is often referred to as the Viking Age.

[20] MR Barnes, 'A nadder/An adder: The nasal shift'. *Neophilologus* V64, Nr 1, 1/80 ISSN 0028-2677.

[21] Nickname explanation; WWW: <http://en.wikipedia.org/wiki/Nickname>.

[22] Report of the Committee on Legal Bibliography (1914) note 52, at 57.

Commercial publishers had started producing popular names tables earlier. By 1924, one could use a popular names table in either the 'Federal Statutes Annotated' or the 'United States Compiled Statutes'. Leaders of the American Association of Law Libraries saw the utility of popular name tables long before the practice was uniform.[23]

The popularity of nicknames for living beings or things is often related to their fame. Such as 'Dubya' for George W. Bush, an exaggeration of Texan pronunciation of 'w', Bush's middle initial; 'Jack The Ripper' for the anonymous murderer of the 1888s, and 'Jack The Dripper' for painter Jackson Pollock who created many of his works by dripping paint over horizontal canvas. 'Gers' or 'Teddy Bears' are the nickname for the Glasgow's Rangers. New York is nicked as 'Big Apple', NYC, and 'The Naked City'. 'The Flying Mouth' is the nickname for the famous lawyer and author of many true crime books F(rancis) Lee Bailey. Together with the other lawyers in O. J. Simpson's murder trial (1994) they acquired the nickname 'Dream Team'.[24]

In the context of information technology, a nickname is a common synonym for a username and is also known as a handle, especially within hacker culture. A nickname in this context is ordinarily associated with any system that requires a login, such as a Website, instant messaging system, or a private network. Such nicknames are routinely employed to enable a certain level of security or anonymity, or for other reasons. The 'Computer Crime Research Center' reported in 2005 that 'Britney Spears' was the most popular hacker nickname.[25] Spears' name is used more often by email spammers than any other moniker. Other top ten's are: Bill Gates, Jennifer Lopez, Shakira, Osama Bin Laden, Michael Jackson, Bill Clinton, Anna Kournikova, Paris Hilton, and Pamela Anderson.

Diminutive Nicknames

Some popular nicknames are diminutive, like girly, hubby, honey, Micky, and Minnie. The most common form is formed by the 'diminutive ending', a syllable tacked on to the end of a name that signifies 'little'. Before the 17th century or so, the most common diminutive endings were the Norman/English 'in' or 'kin' *Jack*, for instance, was originally from the name *Jakin*, a corrupted form of *Jenkin* (John+kin). And the name Hank is short for Han-Kin, or Hen-kin, or Henry-kin. Most of these nicknames have died away, leaving behind only surnames. Only a few of this type of nickname have survived to this day, including: Robin (for Robert) and Colin (for Nicolas).

But in modern-day English, the most common type of diminutive is formed by using the ubiquitous Scottish "ie" (or "ee" or "ey") diminutive ending. It was applied at first, only to names popular in Scotland. Christie was originally a male name, from Christopher, as was Jamie from James, Charlie from Charles, Davey from David, etc. Later, the Scottish "ie" spread to the rest of England. Thus we get Johnny from John, Gracie from Grace, Rosie from Rose, Marty from Martin, doggy from dog, or horsey from horse. A lot more nicknames can be found at Edgar's Name Pages. [26]
Such a diminutive ending exists in English with popular no-surname nicknames like 'boyz', 'girlz' (also: 'grrrlz'), 'ladiez', 'gamez', 'appz', 'warez', by which the users achieve distinguishing identity characteristics (the 'z' means: 'outlaw'). Other languages also know this kind of diminutive endings (e.g. in Dutch: 'jah' for 'ja', 'duh', 'goh', 'denkuh' for 'denken'). Another type of playing with characters is visible in the nickname 'SwEeT AnGeL'. The use of emoticon activating codes, such as '(k)', in a nickname displays an interactive effect at the terminating computer screen. [27]

---

[23] *Law Books and their use, giving as examples of statutes cited by popular name the Sherman Act and the Carmack, Amendment* (2nd edn 1924) p30.

Federal Statutes Annotated was published by Edward Thompson Company in 1906 and 1916, and United States Compiled Statutes was published by the West Publishing Company in 1916. One wonders if these were based on a twenty-page booklet titled 'Popular Names of Federal Statutes' that was published by the Library of Congress in 1923. By 1930, the 'United States Code Annotated' had entered the field with such a table.

The Frank Shepard Company began publication of a pamphlet with the popular names of federal statutes in the late 1920s. Cited from Mary Whisner's 'What's in a Statute Name?'; WWW: <http://www.aallnet.org/products/pub_llj_v97n01/2005-09.pdf>.

[24] Johnnie L. Cochran, Jr. led the winning team of lawyers in the "trial of the century" and in the process became arguably the most famous lawyer in the world. Cochran's successful defence of former football great O. J. Simpson against charges of murder in the televised trial was followed by millions of Americans. Although his trial tactics are still sparking debate, his legal acumen and ability to sway a jury have characterized his legal career.

[25] NN, 'Britney Spears Most Popular Hacker Nickname'. (2005) Source: The Computer Crime Research Center; WWW: <http://www.crime-research.org/news/18.06.2005/1307/>.

[26] Edgar's Name Pages. WWW: <http://www.geocities.com/edgarbook/names/other/nicknames.html>.

[27] An emoticon is a small piece of specialised ASCII art (usually two to five characters, always on a single line, used in text messages as informal mark-up to indicate emotions such as " :-) " = ☺ (smile). There are advanced emoticon generating effects in MSN Messenger.
Concise Oxford English Dictionary, *text message abbreviations and emoticons*; Oxford University Press, 11Rev Ed (8 Jul 2004).

Identity and Body

In the past it was discussed by many philosophers (like Descartes and Locke) to what extent identity is locked to a body. Even before Descartes, it was taken for granted that there was a formal, two-fold distinction between a man's immortal soul and his body, which is 'comprised of flesh, susceptible to temptation and decay.' The English philosopher John Locke did not tie our personal identity to either bodies or souls. [28] In Book II of his Essay, Locke offers an account of identity that was quite different from the common understanding of identity existing in his day which distinguishes between body and soul. He offers a three-fold distinction between the identity of a mass of matter, an organism, and a person. Locke holds both that 'one thing cannot have two beginnings of existence, nor two things one beginning'; and because of that he defines the body as 'two or more atoms joined together in the same mass,' to be incredibly fleeting. If a single atom is either added to or removed from a body then 'it is no longer the same mass, or the same body'. The identity of organisms, including humans, is obviously not so fleeting, as it is defined by 'nothing but a participation of the same continued Life, by constantly fleeting Particles of Matter, in succession vitally united to the same organized body'. Locke not only distinguishes the identity of a 'person' from that of a 'man'; he also argues that personal identity consists not in a continuity of metabolic processes (as does the identity of plants and 'mere brutes'). Because the capacity to think distinguishes a person properly from a brute by, 'her consciousness always accompanies thinking, and 'tis that, that makes every one to be, what he calls self, and thereby distinguishes himself from all other thinking things, in this alone consists Personal Identity' (...)' And as far as this consciousness can be extended backwards to any past Action of Thought, so far reaches the identity of that Person'.

Samuel Drew – after joining the Methodist society at St. Austell in 1785 – studied the relation between personal identity and resurrection [29], and stated that it must be supposed that the principle of identity must have a being: 'We must be satisfied that no body can exist without this principle; however we may differ about its manner of existence, and its constituent parts. But it is absurd to suppose, that the identity of the human body can exist before the body itself is actually called into being. Because, if we could imagine that those radical principles, which constitute the identity of the body, could exist prior to the body, it must be the identity of a body which has no existence. It must, in this case, be the identity of a nonentity.' (...) 'It will therefore follow, henceforth, that no principle of identity can exist as such, antecedent to the union of those numerical parts, of which the body is composed, and from which its existence is always denominated in popular language. And, as bodies have not always had this formal existence; so, neither could this principle of identity, which must be lodged in some secret recess within its confines.' [30] Drew's conclusions 'The soul is a simple, immaterial substance' (...) 'The body is not the man.' (...) 'The ear is no part of the man. The eye is no part of the man; it is only the window of the house.' lay down the foundation of virtual identity and clarify that identity is not tied to the matter. [30]

Merricks explains the different ways in which we may say that a person 'is the same person' as oneself in the past. [31] He uses the example of himself at the age of one. If "person" is defined as a set of characteristics, then in many ways he is not the same person as he was then, as his characteristics are now radically different from a one-year-old child. Likewise we might make the comment 'she is not the same person she was before she became famous' when a person who was once warm and friendly now has become distant and aloof. In other words, "the same as" is being used here to mean 'having the same characteristics as'. But numerical identity is something different from this. For instance, suppose the prosecuting attorney asks you in court whether the man being tried is the

---

Richard Ball, co-owner of an advertising and public relations firm in Worcester, designed the Smiley Face in 1963 to help ease the acrimonious aftermath following the merger of two insurance companies; WWW: <http://elouai.com/smiley-history.php>. The very first emoticon possibly appeared in 1979, first used by Kevin Mackenzie at his typewriter. He is believed to have first used the -) symbol, which meant "tongue in cheek". The technique didn't appear to catch on, and it remained for another to start the fad. At September 19th,1982 emoticons are believed to have been invented (or at least they took hold of the popular imagination) by Scott Fahlman on the CMU bulletin board system. This board was similar in concept to today's newsgroups or message boards, and was intended as a place for people to chat and have discussions. Scott noticed a problem on the board: people would post a humorous comment, but others would not get the joke. This led to countless flames and meaningless discussions. Scott suggested the use of :-) to show pleasure (or indicate a joke) and :-(to show displeasure. This usage caught on like wildfire all over the place, and before long you could see this usage everywhere. WWW: <http://www.cs.cmu.edu/smiley/>.

[28] J Locke (1632-1704) *An Essay Concerning Human Understanding*, ed. Nidditch (Oxford Univ Press 1975), Essay, II.xxvii.1, 3, 6, and 12 from: A Bettesworth, E Parker, J Pemberton, and E Symon, 'The Works of John Locke' (London: 1823) 12 Ed, V 7.

[29] S Drew (1765-1833) An essay on the identity and general resurrection of the human body: in which the evidences in favour of these important subjects are considered, in relation both to philosophy and Scripture. (Publisher: Brooklyn 1811) Sect. II. 5.

[30] S Drew, *Essay on the Immateriality and Immortality of the Soul* (1802), p. 159.

[31] T Merricks, *The Resurrection of the Body and the Life Everlasting*, in: MJ Murray (ed), 'Reason for the Hope Within', (Eerdmans, Grand Rapids 1999), 263.

same person that you saw robbing the bank. It would not do to think to yourself '…well, while robbing the bank he was friendly and approachable, but now he is aloof and distant' and then answer "no". This second kind of identity is numerical identity. According to numerical identity, the woman who is now famous still exists, but she is now in possession of new characteristics.

> 'The man who does not know what death is cannot either know what resurrection is.'
>
> Karl Barth [32]

The identity issues of birth and death, such as mortality and identity – even after death, were popular in the secular theatre of post-Reformation England. People like the idea of returning after their death. They like the mind play with being dead and return, and they use artefacts for the credibility of the resurrection. Theatre plays were and still are a source for such spiritual thought spinning. [33]

Resurrection – by the rise and fall of identities – remained a preoccupation of most self and personal-identity theorists throughout the eighteenth century. Ironically, beginning in the 1960s modern equivalents of resurrection burst back onto centre stage in the debate over personal identity. However, in our own times such supernatural scenarios entered the discussion in the guise of science fiction examples. The play with and the change of identities is handed over to the online environment of the Internet. Resurrection is devaluated to something as "mouse clicking", and can spontaneous happen by using a so-called back-up source to recover the information of a website.

Varying Identities

This particular subtitle points out the relation of role models and the varying of identities. Read how the control of identity leads to the composition of identity standards.

Identity Concept

In the social sciences, identity has recently grown enormously in importance as an explanatory variable and as an independent subject of analysis. However, despite its use as a variable, the definition and measurement of identity is often distressingly vague. Thus, the concept of identity has been defined multiple times in the social sciences and several definitions apply to a range of identity categories, including race, ethnicity, nationality, class, gender, religion and transnational organisations.

The social constitution of identity is defined in types of identities of a person, which are produced permanently through different types of communications respectively social systems: (Somers) [34]
- Interaction system based on presence of people;
- Organisational system based on membership and the communication of decisions;
- Social subsystem based on the communicational 'reachability' by symbolically generalised communication media.

Social and interaction systems are not the same thing. Identity in interaction systems depends on presence and taking part in communication (like speaking and internet chatting), e.g.:
- Congruence of identity: 'I am who I am.' (the way to speak, to walk, the sound of the voice)
- Authenticity: 'I am who I claim I am.' (the way to look like …)
- Identification: 'Who are you? - But I am your old friend! …'
- Authentication: 'Are you who you say you are?'

Given social frames organise spontaneous interactions by giving subjects roughly schematised (but not really standardised) roles that they have to play. Subjects do (or don't) trust each other as subjects, but they cannot trust the frames of interaction (e.g. friendship is not liable under a contract).

---

[32] Karl Barth (1886-1968) was an influential Swiss Reformed Christian theologian. He was also a pastor and one of the leading thinkers in the neo-orthodox movement.

[33] Regarding several Shakespearean moments of resurrection, including the final scene of 'The Winter's Tale' (1610), it can be pointed out that Shakespeare's plays provides some of the best examples of the way Renaissance playwrights worked through the various permutations of the stage resurrection in order to elicit emotional responses from their audiences. Ref 'play-death': K Farrell, Play, Death, and Heroism in Shakespeare (University of North Carolina Press, 1989).

[34] MR Somers, *The narrative constitution of identity: A relational and network approach*, in: C Calhoun, (ed) 'Social theory and the politics of identity' (Basil Blackwell, London 1994) pp 605-649.

There are four main social subsystems in the globalised world, which operate as closed self-organised systems: (Parsons; Luhmann) [35]
- economical system,
- political system,
- law system,
- scientific system.

These systems operate on the base of different "symbolically generalised" communication media: (Küppers) [36]
- payment / non-payment, with prices as condition,
- political power / no power, with programmes as condition,
- right / wrong, with laws as condition,
- true / untrue, with theory and methods as condition.

People can have different and varying identities that influence them in different ways. Generating identities is somehow a psychological necessity of mankind. What does the term identity imply? There is a variety within an identity; it may refer to ethnic, class or religious identities. However, we all have varying identities in the real world, mostly related to a specific role such as being professor, husband, dad, or dance partner. This is not really the same as privacy but we do like to control the knowledge others have of us when we are playing a particular role.

Role Model

Identity is closely linked to one's role, hierarchy, social context and more relating aspects in the environment of a human being. As the environmental conditions change frequently the identity is subject to alterations too.

Identity Standard

Identity is not only related to the role etc. of the individual but can also be understood as the consciousness of certain groups of people to separate themselves from one another and to follow this pattern over a period of time. Identity is therefore deeply influenced by notions of individuality, continuity and consistency.

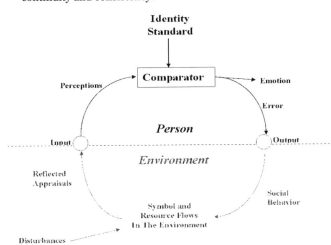

Burke's Identity control theory, with its hierarchical view of identities as control systems, is able to address these issues. [37] Following this identity theory, an identity contains the set of meanings defining who one is in terms of his or her roles (e.g. farmer, wife, or professor), group or social category memberships (e.g. Scot, fraternity member, nerd, or female), or personal characteristics (e.g. dominant, sweet, supportive).

Figure 2: Basic Identity Model (from Burke). An identity consists of the set of perceptions of self-relevant meanings in the interactive situation.

These self-meanings compose what are called identity standards (one standard for each identity). In addition to the identity standard (as shown in figure 2), an identity also consists of the set of perceptions of self-relevant meanings in the interactive situation, and a comparator that compares the perceived self-relevant meanings with the corresponding meanings in the identity standard.

---

[35] N Luhmann, *Social Systems*. (Stanford University Press 1995) p 167 ff, refers:
T Parsons, *Structure of Social Action* [1937] Vol. II. (Free Press 1968).

[36] G Küppers, 'Self-organisation – The Emergence of Order From local interactions to global structures.' WWW: <http://www.uni-bielefeld.de/iwt/kueppers/sein/paperno2.pdf>.

[37] PJ Burke, *Identity Control Theory*, in: George Ritzer (ed) 'Blackwell Encyclopedia of Sociology' (Oxford, Blackwell 2006). WWW: <http://wat2146.ucr.edu/papers/05d.pdf>.

Identity Change

After research in 'real life' Burke[38] concludes that while identities influence the way in which a role is played out, discrepancies between the meanings of the identity standard and the meanings of the role performance will result in change, not only in the role performance (to counteract the discrepancy), but due to the hierarchical structure of identity systems. Change will also occur in the meanings of the identity standard over time (to bring them more into line with the disturbance).

A second mechanism of change in identities is hypothesised to result from persons holding multiple identities that share meanings. Identities that share dimensions of meaning influence each other's standard to maintain the shared meaning at a common level. Changing identity standards redefines who one is, argues Burke. The question of how identities change has been a topic of theoretical interest for a number of years (Burke and Cast; Deaux; Gecas and Mortimer; Kiecolt; McNulty and Swann; Serpe)[39]. Popularized by Erikson as a central psychoanalytic concept, identity is a characteristic defining one's sense of self.[40] That identity change has not been in dispute, but the demonstration of the theoretical mechanisms involved in this change has not been resolved fully, in part because such mechanisms must account for both the stability and the change of identities over time. Burke argues that in this way, every new identity one takes on, through role acquisition or membership in new groups, creates potential changes in other identities that may share dimensions of meaning.

Proceeding, Capurro and Pingel[41] differentiate two kinds of (basic) identity:
- Identity as a metaphysical concept refers to what steadily remains in its appearance. Identity in this sense is granted, according to Aristotle, by substance, i.e. what is supporting changing qualities or accidents and, according to Plato, by the divine model or exemplar (idea).
- Identity as an ontological concept means the possibility of projecting or casting one's life within different existential possibilities. This concept is a richer one since it allows us to relate to different possibilities as such without levelling them out.

Identity and Imagination

This subtitle points out the relations and comparisons between imagination and identity, especially in fantasy and cybernetic worlds, from illusion to virtual reality.

Identity and Inter Reality

Buddhism has long posited that reality is illusory, as in a dream.[42] It is discussed that the concept of the unreality of 'reality' is confusing and not truly accurate because, in Buddhism, the perceived reality is considered illusory not in the sense that reality is a fantasy or unreal, but that our perceptions and preconceptions mislead us to believe that we are separate from the elements that we are made of. In Buddhist thinking, reality is described as the manifestation of *karma*, part of the process of impermanence, similar to the Hindu concept of Maya (Sanskrit: *māyā*, from *mā* 'not' and *yā* 'this'), a term describing many things about illusion. In Hinduism, Maya is the phenomenal world of separate objects and people, which creates for some the illusion that it is the only reality.

---

[38] PJ Burke, 'Identity Change' *Social Psychology Quarterly* 69 2006: 81-96. WWW: <http://wat2146.ucr.edu/papers/06a.pdf>.

[39] PJ Burke and AD Cast, 'Stability and Change in the Gender Identities of Newly Married Couples.' *Social Psychology Quarterly* 60 (1997) pp 277–90.
K Deaux, 'Reconstructing Social Identity.' *Personality and Social Psychology* Bull 19(1993) pp 4-12.
V Gecas and JT Mortimer, 'Stability and Change in the Self-Concept From Adolescence to Adulthood.' pp 265–286 in T Honess and K Yardley (eds), 'Self and Identity: Perspectives Across the Lifespan' (Routledge, Kegan Paul, Boston 1987).
KJ Kiecolt, 'Stress and the Decision to Change Oneself: A Theoretical Model.' *Social Psychology Quarterly* 57 (1994) pp 49–63.
KJ Kiecolt, 'Self Change in Social Movements' pp 110-31 in S Stryker, T Owens, and R White (eds), '*Identity, Self, and Social Movements*' (University of Minnesota Press 2000).
SE McNulty and WB Swann, 'Identity Negotiation in Roommate Relationships: The Self as Architect and Consequence of Social Reality' *Journal of Personality and Social Psychology* 67(1994) pp 1012-23.
RT Serpe, 'Stability and Change in Self: A Structural Symbolic Interactionist Explanation' *Social Psychology Quarterly* 50 (1987) pp 44–55.

[40] EH Erikson, 'Identity and the Life Cycle'. Selected Papers (1959) p 153; Also:
EH Erikson, 'Identity crisis in perspective'. In: EH Erikson, *Life history and the historical moment*. (Norton, New York 1975).

[41] R Capurro and C Pingel, *Ethical Issues of Online Communication Research* (2000); WWW: <http://www.nyu.edu/projects/nissenbaum/ethics_cap_full.html>.

[42] Reality in Buddhism ± 600 BC, Wikipedia; WWW: <http://en.wikipedia.org/wiki/Reality_in_Buddhism>.

For the mystics this manifestation is a fleeting reality. Maya is neither true nor untrue.

In Indonesia (notably the islands Java and Bali), shadow-puppet plays are known as *wayang kulit*. [43] In Javanese, *Wayang* means shadow or imagination, while *Kulit* means skin and refers to the leather the puppets are made from. Stories presented are usually mythical and morality tales. There is an educational moral to the plays, which usually portray a battle between good and evil identities with the good always winning and the evil running away (but eventually to return). The Indonesian shadow plays are sometimes considered one of the earliest examples of animation.

The Turkish-Greek tradition of shadow play "*Karagöz*" (black bye) seems to hail from ancient Egypt. The play was used by various audience communities to negotiate and define cultural boundaries and senses of communal identity. In Islamic culture, representation of a human is a difficult and complex issue. Karagöz is often used to anonymously criticise the politic(ian)s with caricatures, humour and pleasure. The "actor" Hacivat is a bureaucrat; "darling" Karagöz is the people's representative.

Figure 3: Karagöz puppets. The power to create communal identity gave the Karagöz plays their vitality and importance within Ottoman and Greek culture.

Photo © made by Emin Senyer.

Plato's Utopia

The Greek poet-philosopher Aristocles [44], son of Ariston, who took the name Plato as his pseudonym, believed each person moves from the visible to the invisible.[45] Imagination is a stage in this process and is not an evil thing unless it is never outgrown. The divided line moves from the least to the most real. It moves from the weakest form of opinion to sure understanding. For Plato, imagination is where images of the objects of the visible world reside. Here is the realm for the utterly unreflective imitation of what is visible. The imagination confuses the speculative with the visible objects of nature.[46]

> '*Remember how in that communion only, beholding beauty with the eye of the mind, he will be enabled to bring forth, not images of beauty, but realities (for he has hold not of an image but of a reality)*'
> Plato, in: Symposium [47]

The concept of relating realities was already presented by Plato in his 'The Republic' [48] dialogs. Before Thomas More created his Utopia in 1516, Plato had discussed in detail the formation of a perfect nation state: Utopia. And while More's utopia meant an imaginary world, Plato's was essentially a potential reality.

In the early seventeenth century, Robert Fludd, an Oxford-educated physician of wide-ranging and esoteric learning, pictured the interior of the brain containing several interlinked souls, including the imaginative soul [49]: 'fantasy or imagination itself,' writes Fludd, 'since it beholds not the true pictures of corporeal or sensory things, but their likenesses and as it were, their shadows.' (Figure 4)

---

[43] The shadow show in China dates back to at least the Song Dynasty (960–1279), and may date substantially earlier. The art form was spread across the Asian continent by Mongols in the 13th century to countries of the Middle East, where it took root in various forms accept the Turkish-Greek tradition of shadow play, called Karagöz, that seems to hail from ancient Egypt. WWW: < http://www.karagoz.net/english/puppet_theatre.htm>.

[44] Aristocles 427-347 BC, as 'Plato'; WWW: <http://en.wikipedia.org/wiki/Plato>.

[45] A Nehamas, *Plato on Imitation and Poetry in Republic* in: 'Plato on Beauty, Wisdom and the Arts' (Totowa 1982, p 47-78); T Gibson, Plato's Banishment of the Poets (2006); WWW: <http//www.utoronto.ca/mcluhan/tsc_plato_banish_poets.htm>.

[46] Translation is as: art (τέχνη), imitation (μίμησις), images (εἴδωλα; εἰκόνες), imagination (εἰκασία), and phantasy (φαντσία).

[47] Plato (360 BC): *Symposium*. Translation: B Jowett; WWW: <http://classics.mit.edu/Plato/symposium.html>.

[48] Plato (360 BC): *The Republic*. Translation: B Jowett; WWW: <http://classics.mit.edu/Plato/republic.html>.

[49] R Fludd, *Vision of the Triple Soul in the Body* (1619) in: R Fludd, *Utriusque Cosmi... historia* (The History ...of this World and the Other) (Oppenheim, 1617–21). The illustration shows an eye in the same position of the imaginative soul, labelled the

Figure 4: '*Oculus imaginationis*' (The Eye of the Imagination) by Fludd.

The inner eye projects fantasy images onto a screen that lies beyond the back of the head.

Samuel Hibbert [50] published in 1824 a book about visions and identities, which included an elaborate foldout chart about dream states, on which he set out a 'Formula of the various comparative Degrees of Faintness, Vividness, or Intensity, supposed to subsist between Sensations and Ideas'. He tabulated eight transitions in his full cycle, ranging from 'perfect sleep' to 'extreme mental excitement,' and graded fifteen different phases in each of them. They start from 'degree of vividness at which consciousness begins' where it is still possible to impose the will on vision, to 'Intense excitements of the mind necessary for the production of spectres'.

As Frederic Passy said when he founded the Inter-Parliamentary Union in 1889, 'The world is made of achieved utopias. Today's Utopia is tomorrow's reality.' [51] Thomas More might have chosen the literary device of describing an ideal, imaginary island nation with an open political system primarily as a vehicle for discussing controversial political matters freely. Three centuries later, George Orwell did the same in his novel '1984' which discusses privacy and state-security issues from the view of a dystopia. [52]

Danielle Hervieu-Léger [53] points out that the hypothesis that this ever increasing – in view of the constant speeding up of knowledge and technology – Utopian space becomes the space within which religious representations are constantly reorganised. Nevertheless such religious representations are subject to an equally permanent destruction by the forces of rationalism. This tradition evokes a community as a concrete social group or by an imaginary genealogy. The chain which binds the believer to the community and the tradition legitimating this religious belief is what Hervieu-Léger considers as the essential point for imagination.

In 1940 the Argentinean writer and poet Jorge Luis Borges wrote '*Tlön, Uqbar, Orbis Tertius*'.[54] In this short story Borges wrote about the phenomenon of an imaginary world. People imagine (and thereby create) a world that has multiple relations with the real world. One of the major themes of '*Tlön, Uqbar, Orbis Tertius*' is that ideas ultimately manifest themselves in the physical world. There is not a more powerful force than an idea whose time has come. 'It is probable,' writes Borges, 'that these erasures were in keeping with the plan of projecting a world which would not be too incompatible with the real world.'

'oculus imaginationis'. This 'eye of the imagination' radiates a tableau of images: thought-pictures or phantasmata, as Aristotle called them, envisioned by human consciousness. This inner eye, figured by Fludd, does not receive images: it projects them onto a screen that lies beyond the back of the head, floating in a space that does not exist except in fantasy. WWW: <www.blurofotheotherworldly.com/essay_warner.html>.

[50] S Hibbert, *Sketches of the Philosophy of Apparitions* (Edinburgh, 2nd edition 1825; New York: Arno Press, 1975). Samuel Hibbert (1782–1848) was not a singular man of his time; his efforts reflect a pervasive interest in the mind's workings, with numerous counterparts in the literature of psychology, biology, medicine, and literature. His book created a stir because he went on to argue that stomach disorders were a chief source of hallucination, visions, déjà-vu, nightmares, and even mental distress, and that this physical organ could overbear the mind to the point of altering a person's identity.

[51] Inter-Parliamentary Union (IPU); WWW: <http://www.ipu.org/>.

[52] "Nineteen Eighty-Four" (commonly written as *1984*) is a dystopian novel by the English writer George Orwell, published in 1949. The book tells the story of Winston Smith and his degradation by the totalitarian state in which he lives. Along with Aldous Huxley's Brave New World, it is among the most famous and cited dystopias in literature. "Nineteen Eighty-Four", its terminology and its author have become bywords when discussing privacy and state-security issues.

[53] D Hervieu-Léger, *Vers un nouveau christianisme?* (1986), and *La Religion pour Mémoire?* (1993), Translated: Religion as a Chain of Memory. (Rutgers University Press, New Brunswick, NJ 2000).

[54] Jorge Luis Borges wrote *Tlön, Uqbar, Orbis Tertius* around 1940 and it appeared in the Argentine journal *Sur*, May 1940. The story has a postscript dated 1941: Ficciones. (Ed Sur, BsAs 1944) Transl. to English in: JL Borges, *Book of Imaginary Beings*. Emir Rodriguez Monegal reports (in: *Jorge Luis Borges, a Literary Biography*. Dutton, New York 1978) that the idea of "interreality" was used by Borges in 1940 for the phenomenon of an imaginary world.

AM Parker, 'Drawing Borges: A Two-Part Invention on the Labyrinths of Jorge Luis Borges and M.C. Escher' (2001); WWW: <http://rmmla.wsu.edu/ereview/55.2/pdfs/55-2-2001AParkerA.pdf>.

There are several levels of reality (or unreality) in the story:
- Most of the people mentioned in the story are real, but the events in which they are involved are mostly fictional, as are some of the works attributed to them.
- The main portion of the story is a fiction set in a naturalistic world; in the postscript, magical elements have entered the narrator's world.
- The script describes real and fictional people, and real and fictional places. Borges stretches in 'Tlön, Uqbar, Orbis Tertius,' the thin boundaries between reality and unreality to the point where they cease to exist ... if they ever really existed. The idea of "imaginary identity" and "Interreality" was presented.

Interreality and Illusion

Following industrialisation and the civil rights movement, superheroes faced a steady decline in popularity. This was exposed in DC Comics, a leading medium in the US. [55]
The 'Justice League (of America)' or JLA for short is a fictional superhero team. In most incarnations, its roster includes DC's most popular characters.

Comic hero Superman was born in DC's Action Comics #1. In the same issue 'Zatara the Stage Magician' (created as – imaginary – illusionist by Fred Guardineer) imagined a 'green six-sided sun' (jewel) and (King) Ra. The concept of an imaginary world was related from the comics 'Zatara the Magician' (June, 1938 until 1951). As told in the biography and Don Markstein's Toonopedia, the (imaginary) Giovanni 'John' Zatara was constantly merging universes and realities with the personages. [56] For instance: Ccomic characters Zatara and Sindella met each other at the Homo magi race in Turkey and nine months later their daughter Zatanna was born in the USA. This kind of jumping between brain universes also indicates more fusion between the 2D and 3D universes. Zatara returns 2 February 1993 (Season 1, Episode 50) in the TV series 'Batman: The Animated Series' with Zatanna as a stage magician girl friend.[57]

Figure 5: Zatara the Stage Magician (#14, 1939; ©DC Comics)

Together with the comics the Interreality idea was becoming popular at the beginning of the 1990's when computers and Virtual Reality (VR) were coming up and the so called 'cyberspace' (p 42) was booming. Tyler Steele was a character in the Virtual Reality Troopers television series (from 1994) [58] and the co-developer of 'inter-reality travel' alongside Professor Hart who did experiments in Inter-reality travel, and gained the incredible powers required to access cyberspace.

Some sources are claiming that inter-reality travel is really possible. Genesis P Orridge [59] argues that 'we will become the very substance of hallucination, and thus enter and leave it will be the uncertain principle of all realities, regardless of their location.' According to the game environment 'Nexus The Infinite City'[60] this travel is only possible between imaginary realities.

Interreality Comparisons

Interreality is a mix of the virtual and physical realities into a hybrid total experience, and since the pervasion of cybernetics the inter reality comparisons are developing, and step by step it results in a seamless merge of realities. (Kokswijk) [61]

Ascott's thesis on the cybernetic vision in the arts, 'Behaviourist Art and the Cybernetic Vision' [62] begins with the premise that interactive art must free itself from the modernist ideal of the 'perfect object'.[63] Ascott explains in 'The Death of Artifice and the Birth of Artificial Life'[64] that young

---

[55] DC Comics is one of the largest American companies in comic book and related media publishing. (Time -> Warner Bros).
[56] Toonopedia; WWW: <http://www.toonopedia.com/zatara.htm>.
[57] Batman, Zatanna (1993); WWW: <http://www.imdb.com/title/tt0519649/>.
[58] Troopers (1994); WWW: <http://www.imdb.com/title/tt0108978/>.
[59] Genesis P Orridge; WWW: <http://www.mothernyc.com/verbal/onthe.html>.
[60] Genesis P Orridge, 'Nexus The Infinite City' (2001); WWW: <http://www.idiom.com/~trip/gaming/nexus/spd/portability.html>.
[61] J van Kokswijk, *Architectuur van een Cybercultuur* (Bergboek, Zwolle 2003) .
[62] R Ascott, 'Behaviourist Art and the Cybernetic Vision', Cybernetica IX, No. 4 (1966) and X, No. 1 (1967), p.11.
[63] See for 'Cybernetic' Wiener's cybernetics in Chapter 2, page 39.

people today are immersed within real life (RL), virtual reality (VR) or inter-reality (IR). In the 'Museum of the Third Kind' (by Ascott)[65] the hybrid viewer as user or consumer of this art is positioned, bionic to a degree, gender-free, wholly integrated into cyberspace, transculturally oriented to the Net, living globally in the 'interreality' between the actual and the virtual. Post-biological technologies enable us to become directly involved in our own transformation and in changing our being. The emergent faculty of cyberception, our artificially enhanced interactions of perception and cognition, involves the transpersonal technology of global networks and cyber-media. Ascott points out that cyberception not only implies a new body and a new consciousness, but a redefinition of how we might live together in the interspace between the virtual and the real. [66]

Comparing inter realities as research method is first utilised in 1998 by Sorenson, Manz, and Berk [67]. Inter reality comparisons between TV news and crime are reported in 2000 by Dixon and Linz [68] as stereotyping in television. Effects of media stereotypes for inter reality comparison by presence of sex role stereotypes are reported in research of Saintmary's University.[69]
Research results with comparing incidents in real and in virtual worlds are not known yet.

Pseudo Identity

In this subchapter the idea of manipulating the identity is discussed. The value of anonymity is pointed out to be essential for exploring your identity.

Boosting the Anonymity

In the past the idea behind a pseudo identity (pseudonym) was to hide your appearance in order to lower the barrier in contacts between two persons, e.g. love affairs. When the receiver of a message doesn't (need to) know who the sender is, there will be – in most situations – an open attitude and access. [70] To level the interpersonal barriers, e.g. at parties, balls and carnival, masks are used to hide your identity and to boost the anonymity. The mask – already used in the ancient Greek plays to express pseudo identities – is a global metaphor for the loss of identity and recognition.

Masked Identity

The earliest recollection of the use of masks dates back to prehistoric times in relation to the hunting of wild animals. It is assumed that in the Early Stone Age the hunters wore animal-such as masks in order to disguise themselves, making the hunt easier and more cunning. In ancient history masks were used to adapt and express identities of prodigies and divined personalities, such as dragons and gods. In the mystery and miracle plays of medieval Europe, masks were used to portray dragons, monsters, as well as allegorical characters.
In the Renaissance masking was popular in theatre, especially in drama and (light) opera. [71] Still, the carnival festivities in the real world have much in common with the online environment. People wear masks in order to be themselves and to escape the regular social control. The celebration of carnival is a social understanding to express your fantasy, to experiment with sexuality, to change your real age, to make yourself more beautiful, and to try out extravagant behaviour. Much of these manifestations can be found in the online behaviour on the Internet.

Simon Biggs points out that masquerade and hidden identity are increasingly being used to examine the relationship between social, personal and bodily potential in later life. As a key to this

---

[64] R Ascott, 'The Death of Artifice and the Birth of Artificial Life' (1994); WWW: <http://www.phil.uni-sb.de/projekte/HBKS/TightRope/issue.1/texte/royascott_eng.html>.

[65] 'Museum of the Future'; WWW: <http://www.ntticc.or.jp/pub/ic_mag/ic015/ascott/ascott_6_e.html>.

[66] R Ascott, 'The Architecture of Cyberception', Leonardo Electronic Almanac, Volume 2, N 8, *MIT Press Journals*, August 1994

[67] SB Sorenson, JG Manz, and RA Berk, 'News media coverage and the epidemiology of homicide'. *American Journal of Public Health* (1998), 88, pp 1510-1514.

[68] TL Dixon and D Linz, 'Stereotyping in Television'; in *Communication Research*, 27 (5) (2001) pp 547-573.

[69] Saintmary's University, *Effects of Media Stereotypes for Interreality Comparison by Presence of Sex Role Stereotypes* (2006) Third Annual Criminal Justice Conference, Women and the Criminal Justice System: An Agenda for Change, October 28, 2005; WWW: <http://www.saintmarys.edu/~aplamond/c330/stereotypes.html>.

[70] E.g. when you don't want to be identified as the source of the news: 'Your wife is having an affair, (signed) a well wisher'.

[71] The History of Masks in Theatre; WWW <http://www.usq.edu.au/performancecentre/education/ goodwomanofszechwan/masks.htm>.

development, a mask motif has been employed to interpret the management of an ageing self in an uncertain world. The nature of uncertainty, brought about, in part, by the erosion of traditional role expectations and the advent of a consumerist culture, contains elements that are simultaneously threatening and encouraging to experiments with identity in later life. [72]

Switching Identities

When observing the behavioural and explicit experiences of Internet users, you will notice that the young, in particular those less than 25 years old, seem to effortlessly switch between the physical and virtual realities. Through various (often non-converging) media they are supplied with information and contacts. They share address books, lend each other's online identities, make appointments and within seconds hop from one reality to the other. They actually do see a hybrid reality: the crocodile that was first sitting under the bed is yawning with boredom in Sesame Street [73], only to surface seconds later in Digital City foaming at the mouth. In the other window you have to find the hidden salt-water crocodile in the Zoo to lock him up together with the lion before you can free the Allosaurus.[74] If after finding him you quickly vote for this crocodile you earn bonus points. Meanwhile you half and half watch a documentary on the Discovery Channel in which the crocodile grabs a wading aardvark and wolfs it down. A child's magical world is no longer an unstable world, but from the cradle the child feels physically and virtually its way towards the reason that makes 'the world' objective. It discovers that the crocodile has more identities and can evoke both fear and affection; that real teeth look dangerous but that you can crawl through virtual teeth to take an exciting look in the crocodile's stomach.

According to Rushkoff [75] children are also much more challenged to constantly and flexibly deal with changes. Not only do they stand live in front of the boring crocodile enclosure at the zoo while at the same time being busy capturing (virtual) crocodiles on their Gameboy [76], they are also forced to tease and please to get and keep getting attention. They act as Teletubbies and adjust themselves to each environment. [77] They hop from 'real' to 'cyber' and vice versa, and seem to be able to call up a sixth sense such as 'Gestalten'. (Ruesch & Bateson) [78] In the 1960s Bateson calls for clarification of distinction between 'analogue' and 'digital'.[79] Hence, Rushkoff calls it 'the fall of mechanism and the rise of animism' in an era in which everything is self-validating.

Over the years children have noticeably had some difficulty with this 'identity switching'. In both chats and songs the words 'I am real' were frequently uttered. Just as in the ending of each Scooby Doo episode children like to pull off the monster's mask to reveal his true identity and appearance. [80]

Figure 6: Scooby Doo mask, used by the monsters in the TV series.

Illusory Worlds

Since the origin of man people create imaginary worlds with fictitious personages. Mostly in fantasy, but with masks, settings and costumes as well as real scenery. The unreal fiction can be so realistic that people project themselves into those illusory worlds. One of the most famous is Tolkien's

---

[72] S Biggs, 'Choosing Not To Be Old? Masks, Bodies and Identity Management in Later Life' in: *Ageing & Society*, 1997/17: (Cambridge Univ. Press, 1997) pp 553-570.

[73] Worldwide popular American educational children's television series for preschoolers and a pioneer of the contemporary educational television standard, combining both education and entertainment; since 1969; WWW: <http://imdb.com/title/tt0063951/>.

[74] Artificial animals in Zoo Tycoon, a popular simulation computer game. It is a Tycoon computer game, in which the player must run a business, in this case a zoo, and try to make a profit. WWW: <http://en.wikipedia.org/wiki/Zoo_Tycoon>.

[75] D Rushkoff, *Playing the future, what we can learn from digital kids* (Riverhead Trade 1999) p 154.

[76] A handheld mini computer, basically made for playing computer games.

[77] The Teletubbies have the body proportions, behaviour and language of toddlers. They can transform themselves to the context of a situation; WWW: <http://www.imdb.com/title/tt0142055/>.

[78] J Ruesch and G Bateson, *Communication: The Social Matrix of Psychiatry* [1951] (Norton & Co Inc; Reissue edn 1987) p 169.

[79] 'Contesting for the Body of Information: The Macy Conferences on Cybernetics' Lawrence S. Bale, 'Gregory Bateson, Cybernetics, and the social/behavioral sciences'. Published in: *Cybernetics & Human Knowing: A Journal of Second Order Cybernetics & Cyber-Semiotics*. WWW: <http://www.narberthpa.com/Bale/lsbale_dop/cybernet.htm>.

[80] Scooby Doo, famous animated television series since 1969. WWW: <http://www.imdb.com/title/tt0063950/>. The five characters, including the dog Scooby Doo, solve mysteries typically involving tales of ghost, driving around the world in a van called the 'Mystery Machine'. At the end of each episode, the supernatural forces turn out to have a rational explanation (usually a criminal of some sort trying to scare people away so that they can commit crimes).
Source figure: WWW: <http://www.pinatasusa.com/images/scoby%20mascara.jpg>.

imagined world called Arda, and Middle-earth in particular, which were loosely identified as an "alternative" remote part of our own world. Well known since the 1960s are the worldwide Tolkien societies and fan clubs that impassioned followers who consider themselves as Arda inhabitants. [81]

An interactive example of imagination is a telephone chat, in which both participants can call up an illusory world in which they both act. Nicholson Baker's novel 'Vox' [82] spins out the story of how Jim and Abby meet over the phone when they both dial one of those 976 party lines that are advertised in adult magazines. After some exploratory small talk, they retire to the electronic "back room" for a more intimate chat. Not only is there no physical contact, the participants never leave the privacy of their own homes, but still enjoy their common virtual world.

The Value of Anonymity

An imaginary/virtual, identity enables the behavioural role of naming via pseudonyms or hiding personal information by anonymity, as well as the creation and recreation of identities in the computer-mediated social space. Anonymity can be seen as a positive value, when it creates opportunities to invent alternative versions of one's self. Anonymity in online contacts is often used to call in psychological assistance. Using a virtual identity encourages frank response or unbiased exchange in educational and business applications. Furthermore, the use of pseudonyms in intimate personal contacts can be helpful to get to know each other.

On the other hand, Boudourides [83] argues, the use of anonymity or pseudonymity hides identity for the purpose of a decrease in social inhibition and an increase in flaming. [84] Because of many emotional and psychological dynamics, people can be reluctant to interact online. Their faceted identity results in multiple anonymous personas.

Some are defining "identity" as a person, an individual, a constant that cannot change (validated by DNA) and "persona" as an aspect of identity in a specific situation: office persona, parenting persona and so on. They also define "role" as a specific application within a persona. In an office, for example, you might have a manager role, a mentor role, an employee role, etc. This seems to be done to fit all identity, persona, role and their relationship in the identity management hierarchy. The presumed lack of physical and social contextual cues results in several other implications (Cheseboro et al [85]; Kiesler et al [86]).

'Interactants' gain greater social anonymity, because their gender, race, rank, physical appearance, and other features of public identity and indicators of vertical hierarchy, status, and power are not immediately evident (as they cannot be transmitted via computerised text). The disappearance of status and position cues may have a potentially positive effect on group behaviour and group identity. As Kiesler et al note, 'software for electronic communication is blind with respect to the vertical hierarchy in social relationships and organisations'. Participation appears to proceed more evenly distributed across group members. Baron[87] claims that computer-mediated communication makes it difficult for people to dominate and impose their views on others, thus favouring women and minorities. However, Jaffe found that 'women tended to mask their gender with their pseudonym choice while males did not,' an observation underscoring 'the implicit social pressure that women feel when interacting in mixed-gender situations.' (Jaffe et al) [88]

---

[81] John RR Tolkien (1892-1973) was an English philologist, writer and university professor who is best known as the author of The Hobbit and The Lord of the Rings. Tolkien's published fiction includes The Silmarillion and other books, which taken together is a connected body of tales, fictional histories, invented languages, and literary essays about an imagined world called Arda, and Middle-earth (derived from the Old English word "middangeard", the lands inhabitable by humans) in particular, loosely identified as an 'alternative' remote past of our own world. WWW: <http://en.wikipedia.org/wiki/J._R._R._Tolkien>.

[82] N Baker, Vox (Vintage; Reissue edn 1993).

[83] MA Boudourides, 'Social and psychological effects in Computer-Mediated Communication'. (Paper at the 2nd Workshop Conference 'Neties' 1995) WWW: <http://www.math.upatras.gr/~mboudour/articles/csi.html>.

[84] Flaming is the act of sending or posting hostile and insulting messages, usually in online context.

[85] JW Cheseboro and DG Bonsall, Computer-Mediated Communication: Human Relationships in a Computerised World. (Univ. of Alabama Press, Tuscaloosa 1989).

[86] S Kiesler, J Siegel, and TW McGuire, 'Social psychological aspects of computer-mediated communication' American Psychologist, 39(10) (1984) pp 1123-1134.

[87] NS Baron, 'Computer mediated communication as a force in language change' Visible Language, 18(2) (1984) pp 118-141.

[88] JM Jaffe, Y-E Lee, L Huang, and H Oshagan, 'Gender, Pseudonyms, and CMC: Masking Identities and Baring Souls' (Paper submitted for presentation to the 45th Annual Conference of the International Communication Association, 1995) Albuquerque; WWW: <http://members.kr.inter.net/yesunny/genderps.html>.

Sick or Side?

People who frequently practise multiple identities have been diagnosed for years as being mentally sick persons. The usual diagnosis is "dissociation", a psychological state or condition in which certain thoughts, emotions, sensations, or memories are separated from the rest of the psyche. Dissociation most often makes the news with regard to soldiers' responses to wartime stress, rape victims with amnesia regarding the details, and in occasional criminal trials where it is questioned whether a person with Dissociative Identity Disorder (DID) or Multiple Personality Disorder (MPD) can be responsible for his or her actions. [89] Dissociation has a recorded role in murder trials, or at least in movies about murder, where it is occasionally given as a reason for a 'not guilty by reason of insanity' verdict.

Conspicuous is the dissociation 'symptom' of - sometimes addicted - frequent users of the Internet, who bring into being their 'Alter Ego' in online games and societies. Acting with two or more virtual identities, it seems that they also lose control of their identity and questions arise whether such a person can be responsible for his or her actions. As technology enables the management of multiple identities in electronic systems, especially the young generations are entering the era of multi-identity exposure. The reality of this irreality is that at this moment little is known about the long term consequences of this behaviour. Technology developments seem to go much faster than research can follow.

Identity and Identification

The paragraphs in this subtitle are focussed on identifying the identity.

Identity Construction

As expounded in the paragraphs before, the history of modernity has been accompanied by a general weakening of identity, both as a theoretical concept and as a social and cultural reality. This blurring of identity slips downwards to an alarming level in the 20th century. As a theoretical concept, identity nowadays has lost its metaphysical foundation of 'full correspondence' following the destruction of metaphysics by philosophers such as Heidegger, Nietzsche, or Wittgenstein. The "dead god", Nietzsche's metaphor for the demise of metaphysics, has left western cultures with both the liberty of constructing identities, and the structural obligation to do so.
We now face the problem of having to learn how to think without permanent foundations. The new promise of freedom is accompanied by the threat of enslavement. Modern, technologically saturated cultures survive, and act as the gatekeepers of their own technological prisons.
The 'World Information Organisation' [90] argues that for many people identity has become a matter of choice rather than of cultural or biological heritage, although being able to choose may not have been the result of a choice. A large superstructure of purchasable identification objects caters for an audience finding itself propelled into an ever accelerating and vertiginous spiral of identification and estrangement. As the WIO quotes: 'In the supermarket of identities, what is useful and cool today is the waste of tomorrow. What is offered as the latest advance in helping you to "be yourself" is as ephemeral as your identification with it; it is trash in embryonic form.'

Identification by Identity Control

As the Greek play show, people are constantly undergoing changes. Thus it has become difficult to know 'who one is', however, this difficulty is not merely a private problem between two people. To exercise power effectually is also a problem for the state and other institutions of authority as they need to know who you are. With the spread of weak identities, authority is exercised in a different ineffectual manner. Power cannot be enforced without being clear who it addresses. A weakened, hybrid, undefined subject (in the philosophical sense) cannot be a good subject (in the political sense) to an easy subject (in the legal sense) to one's rule. Without adequate identification, power cannot be exercised and dissolved. Identification itself is a necessary precondition for authorita-

---

[89] DSM-IV-TR, *Diagnostic and Statistical Manual of Mental Disorders* Fourth Edition, Text Revision DOI: 10.1176, Appi books.

[90] World-Information.Org is a collaborative effort of organizations and individuals who are directly concerned with issues of participatory involvement in Information and Communication Technologies, and the Internet as we know it today. WWW: <http://world-information.org/>.

rianism; however it is certainly not a sufficient one. [91] Identities are therefore reconstructed using technologies of identification in order to keep the weakened and hence evasive subjects 'subjected'.

Noiriel [92] argues that, due to a combination of personal identity, identification and the urged notion of 'national identity', the nation building since the early nineteenth century has been shaped by colonisation and immigration.

With the western origin of the passports one can quote the letters of marque that the kings (of England and France) gave the merchants and the representatives in order to protect them from the pirates and the ill-treatment of the foreign authorities. The passport term was used for the first time in a treaty signed between England and Denmark on July 11, 1670. This treaty recognised the passport as proof of nationality. Around 1920, the passport became a universal title of voyage and accepted at the conference of the League of Nations in 1922. Lips et al overview:

The History of the Passport:

'If we look at the history of the passport we can observe what could be called a 'révolution identificatoire' in the public domain of nation states (Torpey [93]: following Noiriel [94]).

Whereas the power to regulate citizen movements used to belong to private institutions like the church, or market institutions like serfdom, national governments succeeded in increasingly gaining authority over activities in which a person's status of national citizenship needed to be confirmed. By issuing official national identification papers like the passport, nation states have established the exclusive right to authorise and regulate the movement of people. As identification papers evolved into an administrative expression of national citizenship, citizens have become dependent on nation states for the possession of an official 'identity' which may significantly shape their access to various spaces and activities.

Interestingly the first passports and passport controls for that matter were not so much used to regulate citizens' access to spaces beyond their home country as we are used to today, but to prevent people from leaving their home territory. Consequently those citizens leaving their Kingdom (for instance under the old regime in France) were required to be in possession of a passport authorising them to do so. The main purpose of these documentary requirements was to forestall any undesired migration to the cities, especially Paris (Torpey, p.21).

When geographically based citizen registrations were created and used for providing the personal details in passports, social distinctions started to be made between true 'citizens' and 'non-citizens', also to look for traitors who would obviously belong to the alien, non-citizen category. At that time the French government for instance decreed the establishment of civil status (l' état civil), which determined that an individual could only exist as a citizen once his or her identity had been registered by the municipal authorities, according to regulations that were the same throughout the national territory. Consequently passport controls to enter countries or districts became more extensive. In the 19th century in Prussia, the practice could be found whereby incoming travellers were provided with a passport from the receiving state rather than by the state of the traveller's origin. These passports were no longer issued by local authorities but by higher-level officials. The foreigners and unknown persons circulating in the country were to be subjected to heightened scrutiny by the Prussian security forces, with the assistance of specific, legally defined intermediaries like landowners, innkeepers and cart-drivers (Torpey, p.60).

In the late 19th century a generally liberal attitude of governments toward freedom of movement could be observed; a development which was stopped in the 20th century by national government's desires to regulate immigration. It also targeted the restricted immigration of specific national groups (e.g. USA) and to stimulate economic opportunities for their own citizens abroad (e.g. Italy), to be able to better protect their country for suspicious people in times of war (e.g. Germany, UK, France), or to have the possibility to track their own nationals for conscription into their armies (e.g. Germany). Generally in the 19th and 20th century we may observe a development towards two models for citizenship attribution and the related issuing of passports to citizens, namely on the basis of "ius soli" ('law of the soil') and "ius sanguinis" ('law of the blood') The latter model had to do with the

---

[91] Identity v Identification; WWW: <http://world-information.org/wio/infostructure/100437611729/100438658075>.

[92] G Noiriel, *The Identification of the Citizen: the Birth of Republican Civil Status in France*, in: J Torpey and J Caplan (eds), 'Documenting Individual Identity. The Development of State Practices in the Modern World' (Princeton University Press, Princeton, 2001) pp. 28-48.

[93] J Torpey, *The Invention of the Passport: Surveillance, Citizenship and the State* (Cambridge University Press, 2000).

[94] G Noiriel, *Etat, nation, immigration. Vers une histoire du pouvoir* (Berlin, Paris 2000).

Read also: C Watner, *The Compulsory Birth and Death Certificate in the United States and A History of the Census* in: C Watner and W McElroy (eds), 'National Identification Systems' (McFarland & Company, Inc., Jefferson, North Carolina 2004) p70 resp. 132.

development of enhanced mobility of citizens beyond the state's territorial boundaries, especially for economic reasons, and the possibility for nation states therefore to continuously keep a relationship with citizens living abroad.' [95]

States have traditionally employed bureaucratic identification techniques and sanctioned those who try to evade the grip of administration. Immigrants sometimes have the privilege of carrying two passports, but carrying several passports spotlights you as a spy or dubious outlaw. Not possessing an 'ID' at all is the fate of millions of refugees fleeing violence or economic destitution. Lack of identification that proves someone belongs to a state is structurally sanctioned by 'placelessness'. [96]

Figure 7: Sixty years ago on the roof of the world – Lhasa – the first Tibetan passport - that was recognized by various nations - was prepared by the Tibetan Government and used by Tsepon Shakabpa, the then Tibet's Secretary of Finance (1930-1950). The identity document is a big spreadsheet of the traditional Tibetan hand-made paper folded and old-looking. The passport was issued by the Kashag (Cabinet of Tibet), Lhasa, on 26th day of the 8th month of Fire Pig Year (Tibetan). The date coincides with October 10, 1947. The passport bears official stamps of recognition by countries like France, India, United Kingdom, United States, Italy, and Switzerland which granted visas and transit permits to Shakabpa. [97]

Biometric Identity Control

The 'technisised' acceleration of societies and the weakening of identities make identification a complicated matter. Bureaucratic identification techniques can be technologically bypassed. Passports and signatures can be forged; data can be recreated and manipulated; even real faces are morphed by surgery to fit the fixed picture on the identification certificate. Traditional bureaucratic methods are slow.

The requirements resulting from these constraints are met by biometric technology. Unlike the body rendered *knowable* in the biomedical sciences, biometrics generates a *readable* body: it transforms the body's surfaces into digital codes and ciphers to be read by a machine. Your iris is read, in the same way that your voice can be printed, your fingerprint can be read, and your behaviour can be patterned, by computers that, in turn, have become 'touch-sensitive', and endowed with seeing, hearing and other sensing capacities.

Van der Ploeg [98] argues that biometrics appears to be not as different as the older and existing forms of establishing and verifying personal identity in the deliverance of all kinds of social services and securing economic exchanges. The practices of requiring birth certificates, passports, identity cards or driving licences, providing signatures, pictures, and data like place of birth and current address, have been around for a long time and similarly serve the purpose of proving that one is who one

---

[95] AMB Lips, JA Taylor, and J Organ, *Identity Management as Public Innovation: Looking Beyond ID Cards and Authentication systems* in: VJJM Bekkers, HPM van Duivenboden, and M Thaens (eds), 'ICT and Public Innovation: assessing the modernisation of public administration' (IOS Press, Amsterdam 2006).

[96] The only time Britain had an identity card system was between 1939 and 1952 (the Second World War). The compulsory issue of identity cards was part of the terms of the National Registration Act 1939, a piece of wartime emergency legislation that received the Royal Assent on 5 September 1939. The Act set up a National Register, containing details of all citizens. National identity cards were then issued to all civilians on it. However, there were also ration cards in the First World War.

[97] P Sharma, 'First Tibetan Passport Found after 15 Years' *Hindustan Times*, April 2, 2004.

[98] I van der Ploeg, 'Written on the Body: Biometrics and Identity', in: *Computers and Society*, March 1999, p. 37-44.

claims to be - that is, a person entitled to the services, benefits or privileges applied for. Such identification practices are based on certified documents issued by certifying agencies and institutions, and subsequent chains of such documents that serve their purpose by virtue of their referring to each other. For example, a birth certificate is needed to get a passport; a passport, in return, is requested for the registration of birth of your child. Referring all events through the identification to a specific person results in a huge amount of information that builds a profile of that person. [99]

Nevertheless, how can a biometric identifier be both ideally identifying and not say anything particular about you? Van der Ploeg considers that the key to this riddle may be found in the idea that meaning is not something intrinsic, but, following the philosopher Wittgenstein, determined by use. Following this kind of reasoning, we should perhaps not expect to be able to determine any intrinsic meaning of biometric data, or the biometric body in general, but investigate quite specifically what uses and practices biometrics will become part of. Hence, biometrics would become one of the clearest examples of the way technology renders the nature-culture distinction and the nature-nurture debate obsolete altogether, since the difference between natural bodies and social structures has become meaningless. Just as our culture of biotechnology transforms innate bodily characteristics, rendering 'nature' more and more an object of design, through biometrics bodies may become inscribed with identities shaped by longstanding social, cultural, and political inequalities.

Summary Chapter 1: **'Identity and Identification in History'**

In this chapter we looked at an overview of identity in the past. Philosophers played the main role in the development of the meaning of personal identity, but game playing Greeks, data mining Incas, dominating Vikings, resurrecting Methodists, Scottish phonemes, popularising lawyers, and even Britney Spears, Osama Bin Laden, Michael Jackson, and Bill Clinton performed their role in the identity change. Generating identities became a psychological necessity of mankind. There is a variety of an identity which may refer to ethnic, class or religious identities. Each identity contains the set of meanings defining who one is in terms of his or her roles (e.g. farmer, wife, or professor), group or social category memberships (e.g. Scot, fraternity member, or female), or personal characteristics (e.g. dominant, sweet, or supportive). These self-meanings compose what are called identity standards (one standard for each identity).

As technology enables the management of such multiple identities in electronic systems, especially the younger generations are entering into the era of multi-identity exposure. The reality of this irreality is that little is known about the long term consequences of this behaviour. Technology developments seem to go much faster than research can follow and already shake at the principles of identity. For years a general weakening of identity, both as a theoretical concept and as a social and cultural reality, blurs the identity downwards to a simple proof of existence. Forged passports and signatures, recreated and manipulated data, and morphed faces, make identification of a human being a complicated matter, which seems to be tackled with a biometric control. The next chapter focuses on identity in the "online" context.

---

[99] Van der Ploeg indicates the example of a person applying for a university student card, who must present a passport to get the university library card, and so on. Thus the right to walk into the library, to make use of its computers, catalogues, attendants' time and expertise, and to take valuable books home, is premised on a set of identity markers that together, and by internal reference, establish that one is student so and so, who paid the university tuition, paid previous fines on late returns, and thus is a deserving member of the population the library is there to serve.

2 The Impact of Technology

'Imagination is more important than knowledge; knowledge is limited, imagination conquers the world'

Albert Einstein (1879-1955)

Introduction
As mentioned earlier, modern telecommunication facilities have allowed an enormous number of people around the world to communicate with each other. Time and space have therefore acquired another dimension and new phenomena come into existence, such as 'virtual experience'. The Internet becomes an extra window from your home (or workplace) that overlooks the outside world. The more often you climb through this imaginary window, the more it becomes a 'door' to the virtual society in the virtual space. Recall from chapter one that the type of number values, appreciations and coupled relations in the quipu cords of the Incas is nowadays in use on the Internet by URL hyperlinks and search engines, this chapter discusses how technology can help us to open the imagination and to close the gap between past and present.

This section focuses to the development of technology to empower the imagination. It gives an overview of the technical background of the virtual identity and the relation with the telecommunication technology; from connection to common use. It headlines topics such as access, infrastructure, identification, security and virtual reality.

Empowering the Imagination
This subtitle points out the influence of new technology on the empowerment of fantasy and imagination. History shows that there are more Internets to deal with.

Reality and Truth
The Internet and telephony lead to the establishment of virtual contacts on a large scale on the basis of interests or involvement, whereby each contact in itself is part of one of the many groups (so-called communities of interest), where practise shows that they are using different kinds of identities. But who is the real human? What is the trustworthy identity? 'Everyone his truth'[100] is a fundamental thought, but 'truth' and 'reality' are not synonymous, neutral, objective terms, but terms that are relative to (1) the group of people and (2) the time in which one lives. Truth is always historical and relational, and reality relates to 'consciousness' and the present.

As early as 1934 Popper wrote of the search for truth as 'one of the strongest motives for scientific discovery'. Still, he describes in 'Objective Knowledge'[101] early concerns about the much-criticised notion of truth as correspondence.

Novelist Winterson opposes the artist and the realist. She maintains that unlike the realist, the artist is able to understand that 'there is more around us than the mundane' ([102] p 136) and that therefore the artist is able to understand reality in a more profound way. By placing truth outside the directly accessible world of the senses, Winterson says in 'Art Objects', her collection of essays:

---

[100] Godfried Bomans (1913-1971), a famous Dutch author, once exclaimed 'Everyone his truth' as a contradiction.

[101] K Popper, *Objective Knowledge: An Evolutionary Approach*, 1972, Rev. ed., 1979.

The semantic theory of truth formulated by the logician Alfred Tarski and published in 1933. Popper writes of learning in 1935 of the consequenoeo of Tarski's theory, to his intense joy. The theory met critical objections to truth as correspondence and thereby rehabilitated it. The theory also seemed to Popper to support metaphysical realism and the regulative idea of a search for truth. According to this theory, the conditions for the truth of a sentence as well as the sentences themselves are part of a metalanguage. So, e.g., the sentence 'Snow is white' is true if and only if snow is white. Although many philosophers have interpreted, and continue to interpret, Tarski's theory as a deflationary theory, Popper refers to it as a theory in which "is true" is replaced with "corresponds to the facts." He bases this interpretation on the fact that examples such as the one described above refer to two things: assertions and the facts to which they refer. WWW: <http://en.wikipedia.org/wiki/Karl_Popper>.

[102] J Winterson, *Art Objects* (Jonathan Cape, London 1995). Jeanette Winterson's subsequent novels explore the boundaries of physicality and the imagination, gender polarities, and sexual identities.

> It is through the painter, writer, composer, who lives more intensely than the rest of us, that we can rediscover the intensity of the physical world. And not only the physical world. There is no limit to new territory. The gate is open. Whether or not we go through is up to us, but to stand mockingly on the threshold, claiming that nothing lies beyond, is something of a flat earth theory. ([103]: p 151)

Whereas Descartes [103] believes that the use of reason alone can provide us with realistic knowledge of the world, Winterson believes that: [104]

> The realist (JvK: from the Latin "res" = thing) who thinks he deals in things and not images and who is suspicious of the abstract and of art, is not the practical man but a man caught in a fantasy of his own unmaking. ([103]: p 143)

Imagination 'Powered by Technology'

Fantasy Empowered

In the 'Renaissance' period the consideration about 'fantasy or imagination' was a very popular theme for philosophers and scientists. Following Robert Fludd's theories about the interior of the brain containing several interlinked souls, including the imaginative soul, the extension of the human senses by technology started in scientific experiments.

First results were shown by the German Jesuit Athanasius Kircher [105] (figure 8). By the flickering light of an oil lamp, Athanasius Kircher projected a series of images engraved on glass onto a wall. He could use his projector to illustrate lectures or simply to amuse his visitors.

> Kircher explained: 'The use of a mirror in this new lantern, though, does not differ from what we demonstrated ... in other ways than that it is used in conjunction with a portable lamp while we use sunlight, reflected in a mirror on which images are painted, to display anything that such a portable lamp can display on a wall inside a room or a house, with natural rendering of the colours. We did also, at the same time, demonstrate methods of exposing pictures without the aid of sunlight, either with a concave mirror or a transparent lens. I have mentioned all these details so that the Reader may understand the origins of this new, mysterious lantern, that, as far as we under-stand, not undeservedly is called "The Magic Lantern" or "The Sorcerers Lamp" due to its remarkable capability to let the vision of any object come to sight in a dark room or in the silence of the night.'

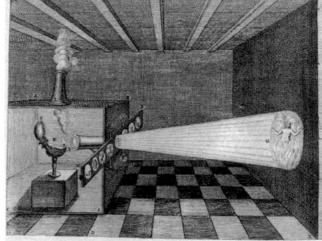

Figure 8: Laterna Magica by Athanasius Kircher (1646)

Kircher's *Musurgia Universali* [106] has been recognised as being a very important step in the history of acoustic theory. The work shows how reverberations and echoes can be bounced for long periods of time in complex wall structures. It's basically a 'piazza-listening device' for the purpose of kings. Kircher explained that 'the voices from the piazza are taken by the horn up through the mouth of the statue in the room on the *piano nobile* above, allowing both espionage and the appearance of a miraculous event'.

---

[103] R Descartes, *Discours de la méthode. Pour bien conduire sa raison et chercher la vérité dans les sciences.* (Garnier-Frères, Paris 1960); Transl. ed Discourse on Method and the Meditations (Penguin Books, London 1968).

[104] A Estor, 'Jeanette Winterson's Enchanted Science' (PhD dissertation Leiden University, Leiden 2004); WWW: <https://openaccess.leidenuniv.nl/bitstream/1887/9924/1/Proefschrift_Estor.pdf>.

[105] Athanasius Kircher (1602-1680) was a 17th century German Jesuit scholar who published around 40 works in the fields of oriental studies, geology and medicine. In: Athanasius Kircher, Ars Magna Lucis Et Umbrae, In X. Libros Digesta, Sumptibus Hermanni Scheus MDCXLVI Romae 1646, p 793. WWW: <http://www.faculty.fairfield.edu/jmac/sj/scientists/kircher.htm>.
TL Hankins and RJ Silverman, *Instruments and the Imagination* (Princeton University Press; 1999).

[106] Source: Kircher exhibition Glasgow 2002; WWW: <http://special.lib.gla.ac.uk/exhibns/month/nov2002.html>.

Figure 9: Athanasius Kircher, in: Musurgia Universali (1650)

From Connection to Common Use
This subchapter recalls the interesting stories about the experiments that lead to today's cyber-reality. Making the invisible visible indicates the importance of technology.

Tele-technology

Shocking Experiments
The forerunner of all electric networks is found in France. On an April day in 1746 at the Grand Convent of the Carthusians in Paris, some two hundred monks arranged themselves in a long, snaking line. Jean-Antoine Nollet, the Abbot of the Grand Convent, decided to test his theory that electricity travelled far and fast. Between each pair of monks a 25-foot iron wire was spun. Once the reverend fathers were properly aligned, Nollet hooked up a battery to the end of the line and noted with satisfaction that all the monks started swearing, contorting, or otherwise reacting simultaneously to the shock. A successful experiment: an electrical signal can travel a mile and it does so quickly.

Télégraphe
From 1753 –'An Expeditious Method of Conveying Intelligence'– until 1837 –'Five Needle Design'– at least 60 experimental electric telegraph concepts are known to have been constructed by a number of researchers, however the word '*télégraphe*' was first used for the semaphore, a kind of optical distance messaging, invented in France in 1791 by the brothers Chappe. Unaware that other inventors had failed after being unable to transmit signals over long wires, Samuel Morse developed in 1838 a long distance telegraph system and created the Morse code, an electronic alphabet that was patented in 1840.
Rapidly a new communications technology was developed that allowed people to communicate almost instantly across great distances, in effect shrinking the world faster and further than ever before. Once more and more telegraph networks were connecting the many parts of the globe, so-called "operators" were employed to connect people from and to connections in one or more networks. Those female telegraphers (today, we call them webmasters) were skilled in sending and receiving the Morse code without using paper tapes. This was the beginning of the transient high-tech workforce.'[107]

The Victorian Internet
This internet inundated its users with a deluge of information and gave rise to new forms of crime. Secret codes were devised by some users, and cracked by others.
Chat rooms and the online social life arose in the 1860s. Telegraph systems were as the spokes of a wheel. Every message that was sent along one spoke was heard by every telegraph operator along

---

[107] A Ramos, 'The Steam-powered Internet'. WWW: <http://www.andreas.com/faq-steamnet.html>.

that spoke. During slow hours, telegraph operators chatted with each other, swapping stories, jokes, news, rumours, and played chess or checkers. Telegraph operators added their initials to the end of transmissions, in order to identify themselves. These signatures (the first electronic virtual identities!) were called "sigs".
All sorts of local abbreviations, similar to today's LOL, BTW, and 69 were invented. Meetings were held online, with hundreds of telegraph operators, strung out along a 700-mile line, in attendance. Friendships arose between people who never met in person. Online romances were common and sometimes, once they had met for the first time, these romances would come to an abrupt end. In 1879 a telegraph operator named Ella Cheever Thayer wrote '*Wired Love*', a novel about an online romance between two telegraph operators [108]. An article entitled 'The Dangers of Wired Love' [109] told the story of George McCutcheon, who installed a telegraph in his newsstand and got Maggie, his 20-year old daughter, to operate it. Soon, she was flirting online with several young men, and shortly after, she was involved with Frank Frisbie, a married man. Father yanked out the telegraph, but Maggie found a job at a nearby telegraph office and resumed the online affair. Marriages were performed online, with the minister telegraphing the ceremony, and the bride and groom, apart in different cities, taping 'I do'. All of the telegraph operators along that line were present online at the wedding. (Ref: Standage) [110] This was the first online community. [111]

The telephone was initially seen as a broadcast medium: '...dancing party (with) ... no need for a musician' (Nature, August 24, 1876) and 'The serenading troubadour can now strum his guitar before the telephone, undisturbed by apprehension of shotguns and bull dogs. Romeo need no longer catch a cold waiting at Juliet's balcony' in: 'A Call' (Ford) [112]. After a short time the telephone made the telegraph personal by connecting home to home. However, the broadcast variant, called rediffusion, still continued until far into the 20th century.

Soon after, in the early 1900s, thanks to the worldwide development of connected electronic networks, out on the wires, a technological subculture with its own customs and vocabulary established itself, and those who were (only) connected by lines created an imaginary world for themselves.

Wireless

' *The wireless telegraph is not difficult to understand. The ordinary telegraph is like a very long cat. You pull the tail in New York, and it meows in Los Angeles. The wireless is the same, only without the cat.* '

Albert Einstein (1897-1955)

First developed in 1879, with the advent of improved technologies such as vacuum tubes and rectifiers, the wireless radio was honed into an interesting little device. Once radio signals could be transmitted and received with improved clarity around 1920, the idea of public radio began to take hold in the western world.

Figure 10: CBC Radio Drama 1932 [113]

---

[108] EC Thayer, *'Wired Love' a Romance of Dots and Dashes* (1879) (WJ Johnston, New York 1880).

[109] NN 'The Dangers of Wired Love.' The story of New York telegrapher Maggie McCutcheon. *Electrical World, Boston Globe*, 13 February 1886, 68-69.

[110] T Standage, *The Victorian Internet. The Remarkable Story of the Telegraph and the Nineteenth Century's Online Pioneers.* (Berkley, New York 1998)

[111] Virtual communities are also created by other media and have existed before cyberspace was discovered. Poster argues that 'just as virtual communities are understood as having the attributes of 'real' communities, so 'real' communities can be seen to depend on the imaginary: what makes a community vital to its members is their treatment of the communications as meaningful and important. Virtual and real communities mirror each other in chiasmic juxtaposition'.
Cited from: M Poster, *Post-modern Virtualities*, in: 'The Second Media Age' (Blackwell Cambridge, 1995).

[112] FM Ford (cited Ford Maddox Ford, but born Ford Madox Hueffer), *A Call: The Tale of Two Passions* (Chatto & Windus 1910).

These quotes cover an interesting point. Traditionally a distinction has been drawn between broadcasting (one to many) and one to one communication. Rediffusion distributed by telephone lines was considered as broadcasting by wire.

[113] Source: NN. WWW: <http://static.flickr.com/92/224918592_b887d3e5c9_o.jpg>.

The first public radio broadcasting stations were an instant success; listeners would sit around the radio listening to everything that was broadcast. As a result many more radio stations popped up during the 1920s, some even overnight. Radio provided a cheap and convenient way of conveying information and ideas. The first broadcasts consisted primarily of news and world affairs. Later on in the decade, radios were used to broadcast everything from concerts to sermons. Soon radio playwrights created an imaginary world inside someone else's head. Unknown 'virtual' actors became popular through their voice animated dramatis personae. The radio not only brought people together (in nations), but it also brought a whole new way for people to communicate and interact.

Invisible Visible

The transmitting of images – to make the invisible visible – also improved fast. From 1831 on, many scientists such as Faraday, Caselli, Goldstein, Bell, Edison, worked simultaneously on the concept of recording, transmitting and reproducing images. Calling it the electric telescope, in 1884 Paul Nipkow sent images over wires using a rotating metal disk technology that had 18 lines of resolution. At the 1900 World's Fair in Paris, the first International Congress of Electricity was held, and the Russian Constantin Perskyi made the first known use of the word 'television'.

Soon after 1900, the momentum shifted from ideas and discussions to the physical development of television systems. [114] Two major paths in the development of a television system were pursued by the inventors. Besides Nipkow other inventors attempted to build electronic television systems based on the cathode ray tube developed independently in 1907. In 1927 Bell Telephone and the U.S. Department of Commerce conducted the first long-distance use of television which took place between Washington D.C. and New York City on April 9th. US Secretary of Commerce Herbert Hoover commented: 'Today we have, in a sense, the transmission of sight for the first time in the world's history. *Human genius* has now destroyed the impediment of distance in a new respect, and in a manner hitherto unknown.' Between 1930 and 1950 the 'television' developed into a public broadcast service.

The next step was to fool the audience in their experience by manipulated imagination. In 1950, Warner Bros employee and ex-Kodak researcher Arthur Widmer began developing blue screen techniques: one of the first films to use them was the 1958 adaptation of the Ernest Hemingway novella, *The Old Man and the Sea*, starring Spencer Tracy. The background footage was shot first and the actor or model was filmed against a blue screen carrying out their actions. To simply place the foreground shot over the back-ground shot would create a ghostly image over a blue-tinged back-ground. The actor or model must be separated from the background and placed into a specially-made "hole" in the background footage. In the 1980s the introduction of digital compositing, "colorkey" [115], enhanced the virtual possibilities. Ever since then the image of a person can be separated from the background (context) so that a person can be made virtually and e.g. transferred to each other context.

Cybernetics

In 1934, decades before Ted Nelson coined the term 'hypertext', Otlet [116] envisioned a new kind of scholar's workstation: a moving desk shaped as a wheel, powered by a network of hinged spokes beneath a series of moving surfaces. The machine would let users search, read and write their way through a vast mechanical database stored on millions of 3×5 index cards. He called the machine Mundaneum.

Vanavar Bush presented in 1945 the "memex", a 'desktop' device in which an individual stores all his books, records, pictures, and communications, and which is mechanised so that it may be consulted with exceeding speed and flexibility. It is an enlarged intimate supplement to his memory. (Figure 11)

---

[114] NN, 'History of Communications', FCC WWW: <http://www.fcc.gov/omd/history/tv/1880-1929.html>.

[115] Color key, also called 'colour key' and 'chroma key', is a technique for superimposing one video image onto another. Widely used to place an interesting scene behind people such as a news reporter on TV, it is also used for creating special effects such as a man walking on the sun or a car floating on the ocean. The foreground image is shot in front of a backdrop, which is a single, solid colour, typically blue or green. When both images are combined, the background pixel takes precedence wherever a pixel of the solid colour is found in the foreground image. If the foreground image is a person and the backdrop is blue, no articles of blue clothing can be worn or the background image will bleed through in those locations. WWW: <http://en.wikipedia.org/wiki/Bluescreen>.

[116] P Otlet; WWW: <http://www.mundaneum.be/> and <http://www.boxesandarrows.com/view/forgotten_forefather_paul_otlet>.

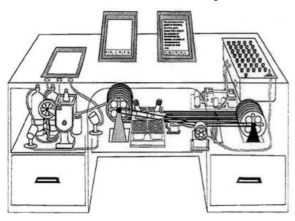

Figure 11: The Memex (Rev.) original source: Life 19(11), p 123.

In 1948 Norbert Wiener wrote a manifesto for a new science, titled 'Cybernetics [117]: Control and Communication in the Animal and the Machine'. In it he gives an overview of a twofold and parallel history, that of the automaton and of the human body. Wiener distinguishes four great periods in the history of the automaton; a mythical Golem era, the era of the clock, the steam era, and the era of control and information. Each one has its own model of the human body. The body as a mouldable figure, as a clock, as a refined steam engine and as an electronic system.

Simultaneous to the development of machines, in which servomechanisms are the most important, the discovery of neurons ensures that neither machine nor human body can be considered as conservative systems with rather limited possibilities. The study of automata 'whether in metal or in flesh' is imbued with terms such as message, noise, coding, information amount and feedback. The image of a communication network springs to mind.

It is the feedback – which implies bidirectionality or interactivity, particularly its increased scale – that forms the core of the cybernetic system. Steam engines, too, had feedback technologies, but the complex technological systems that surfaced in the 1940s allowed a deep penetration in the social field, where previous automata were restricted to the industrial tissue of a nation. To Wiener the application of this new technological theory was not restricted to the world of machines but equally valid for the contemplation of the workings of the human body and mind. Body and machine are no longer considered as systems whose main function or activity is to save or transfer energy. It is communication that enables the functioning of both organism and machine. Concepts such as "cyborg" [118] and cyber-space can be imagined only within the cybernetic and Wiener stands at the origin of these terms.

Cyber-Reality

Virtual reality (VR) [119] is a technology which allows a user to interact with a computer-simulated environment, be it a real or imagined one. Most current virtual reality environments are primarily visual experiences, displayed either on a computer screen or through special stereoscopic displays, but some simulations include additional sensory information, such as sound. In medical and gaming applications some advanced haptic [120] systems now include tactile information, generally known as force feedback. The simulated environment can be similar to the real world, for example, simulations for pilot or combat training, or it can differ significantly from reality, as in VR games. It is unclear exactly where the future of virtual reality is heading. In the short run, the graphics

---

[117] The name cybernetics was coined by Norbert Wiener to denote the study of 'teleological mechanisms' and was popularised through his book Cybernetics, or Control and Communication in the Animal and Machine (1948). The word cybernetics ('cybernétique') had, unbeknownst to Wiener, also been used in 1834 by the physicist André-Marie Ampère (1775–1836) to denote the sciences of government in his classification system of human knowledge. Such a word was also used by Plato in The Laws to signify the governance of people. The words govern and governor are also derived from the same Greek root. Teleological mechanisms (from the Greek τέλος or telos for end, goal, or purpose) in machines with corrective feedback date from as far back as the late 1700s when James Watt's steam engine was equipped with a governor, a centrifugal feedback valve for controlling the speed of the engine. WWW: <http://www.reference.com/browse/wiki/Cybernetics>.

[118] A cyborg is a cybernetic organism, such as an organism that is a self-regulating integration of artificial and natural systems. The term was coined in: M Clynes and N Klime, 'Cyborgs and Space' in: Astronautics (1960) pp 26-27 and 74-75.

[119] The origin of the term virtual reality is uncertain. The VR developer Jaron Lanier claims that he coined the term. A related term by Myron Krueger, 'artificial reality', has been in use since the 1970s. The concept of virtual reality was popularized in mass media by movies such as "Brainstorm" and "The Lawnmower Man", and the VR research boom of the 1990s was motivated in part by the non-fiction book Virtual Reality by Howard Rheingold. The book served to demystify the subject, making it more accessible to less technical researchers and enthusiasts, with an impact similar to what his book The Virtual Community had on virtual community research lines closely related to VR. It has long been feared that Virtual Reality will be the last invention of man, as once simulations become cheaper and more widespread, no one will ever want to leave their 'perfect' fantasies. WWW: <http://en.wikipedia.org/wiki/Virtual_reality>.

[120] Haptic, from the Greek αφή (Haphe), means pertaining to the sense of touch.

displayed in the head mounted display will soon reach a point of near realism. The audio capabilities will move into a new realm of three dimensional sound. This refers to the addition of sound channels both above and below the individual. The virtual reality application of this future technology will most likely be in the form of over-ear headphones. In order to engage the other senses, the brain must be manipulated directly. This would move virtual reality into the realm of simulated reality à la movie 'The Matrix' [121].

Today it is very difficult to create a high-fidelity virtual reality experience due largely to technical limitations on processing power, image resolution and communication bandwidth. However, over time those limitations are expected to be overcome as processor, imaging and data communication technologies become more powerful and cost-effective.
Although no form of this has seriously been developed at this point, Sony has taken the first step. On April 7, 2005, Sony went public with the information that they had filed for and received a patent for the idea of the non-invasive beaming of different frequencies and patterns of ultrasonic waves directly into the brain to recreate all five senses. [122] Research had shown that this is possible, however, Sony has not conducted any tests as of yet and says that it is still only an idea.

Cyberspace

Cyberspace is the location where one stays during a telephone call. (Featherstone and Burrows). [123] However, Gibson abstracts the word to a 'consensual hallucination':

> '*Cyberspace. A consensual hallucination experienced daily by billions of legitimate operators, in every nation, by children being taught mathematical concepts... A graphic representation of data abstracted from banks of every computer in the human system. Unthinkable complexity. Lines of light ranged in the nonspace of the mind, clusters and constellations of data. Such as city lights, receding... Cyberspace is the total interconnectedness of human beings through computers and telecommunication without regard to physical geography.*'   William Gibson 1984 [124]

In the 1970s the US military developed "ARPA net" (as predecessor to the Internet), and extended it to a worldwide scientific network, including email features and the Usenet as a bulletin board. When the Internet was made available to universities and to the public for common use (in 1991), and was transformed into a ubiquitous public network by opening the World Wide Web, it gained a prominent place in our daily communications thanks to easy access (home connection, relatively low costs, browser interface and Web applications). The wireless Internet was introduced at the end of the last century as a concept that combined wired computers and wireless telephones. At first the sometimes addictive wire line phoning and so called 'internetting' usually took place in a secluded and private environment, but now mobile phones with Internet access are freely being used out in the streets, outdoor cafes, trains, and shops.

The telegraph, telemetry, telephony, radio, television and the Internet, together termed "telecommunication", have led to an acceleration of message traffic worldwide. Time zones have fallen away and distance has become a relative concept. Identity frees the human body. Telecommunications extends our senses. Thanks to the telegraph we write out of sight. Due to telemetries we signal from a distance and with the telephone we can have remote online communication. With television we look over the horizon and by radio we listen out of earshot. Vibrations (touch) and aromas (smell) can be remotely captured and (re)produced. With the extension of our senses, we also move and enhance our experience of space and society. With 'feeling' we achieve virtual reality such as being totally absorbed by a book or film. [125] This is not hype, but seems to become a new world: the 'cyberworld', created by and existing by the grace of tele-technology.

The term "cyberworlds" was introduced by Tosiyasu L. Kunii in 1998:

---

[121] The Matrix is a science fiction / action movie. The film describes a future in which the world is actually the Matrix, a simulated reality created by sentient machines in order to pacify, subdue and make use of the human population as an energy source by growing them and connecting them to the Matrix with cybernetic implants. WWW: <http://www.imdb.com/Title?0133093>.

[122] M Horschnell, 'Sony takes 3-D cinema directly to the brain'. *The Times online*; WWW: <http://www.timesonline.co.uk/tol/news/uk/article378077.ece>.

[123] M Featherstone and R Burrows,'Cultures of Technological Embodiment: An Introduction', in: M Featherstone and R Burrows (eds), *Cyberspace, Cyberbodies, Cyberpunk* (Sage, London 1995).

[124] W Gibson, *Neuromancer* (Time Warner International1989).

[125] G Bateson, in an interview with Daniel Goleman. *Psychology Today*, p 44 (1978), cited in: *Nature of Reality* ; WWW: <http://www.afn.org/~gestalt/reality.htm>.

> '*Cyberworlds are information worlds formed in cyberspace. Cyberworlds can be created either intentionally or spontaneously. As information worlds, they can be virtual or real, as well as mixed reality. In terms of information modelling, the theoretical ground for the Cyberworlds is far above the level of integrating spatial and temporal database models. Cyberworlds have been created and applied in such areas as e-business, e-commerce, e-manufacturing, e-learning, and cultural heritage.*'
>
> Tosiyasu L. Kunii [126]

The Next Ten Years

The Bio-PC, our digital brain, is coming. [127] Bio suggests something biological. Indeed. A minuscule programmable biocomputer carries out a billion actions per second within a 99.8% accuracy. Instead of formulas and algorithms to solve a problem the microscopic small computers make use of DNA molecules that store and activate coded information in a living organism. DNA has the potential to work faster than ordinary processors. Nano-scale computers from biomolecules are so small that processes are spread over several cell computers at the same time, as a result of which a trillion computers together can carry out a billion processes. As a result the attention is focused on the technology that links up electronics networks in the environment of people to the human nervous system.

Visionary researcher Ray Kurzweil [128] and MIT's Media Lab founder Negroponte – whose motto is: "Move bits, not atoms." – seem to be sure about the evolution of the cyber-environment. 'Fixed will be mobile, and reversed. Fluid will be solid, and backwards. Organisms will be machines, and the other way round.', they pontificate.

In view of the high realisation potential of the technologies discussed, another stage in the evolution, or in any case a transformation of the current humans, seems to be inevitable. After all, many of the technologies the transhumanists have placed their hopes in, have already become reality (genetic engineering, cloning, implanting peacemakers and artificial joints, heart valves, insulin pumps and electronic senses). They have proven to be successful at least for so called components (artificial intelligent chess programmes), or have at least been successfully tested in laboratories (linking of information transfer between neurons and electronic processors, the nano-technological rearrangement of atoms, successful cryogenic suspension of baboons).

The extrapolation of the current increase in computer processor speed (55-60% per year) – Moore's Law [129] –, linked to predictions of nanotechnologists, makes it probable that the capacity of computers in the near future will be millions times bigger than our current digital housemates. According to trans-humanists it seems that these developments all point to one direction: transformation. In the meanwhile, it is unpredictable whether these developments will lead to the bionic buddy who will organise our lives, to a digital clone of our mind, or at last to a machine-duplicated human.

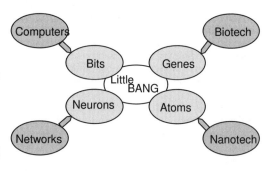

Figure 12: Expected convergence of Bits, Neurons, Atoms and Genes [130]

---

[126] TL Kunii, *The Architecture of Synthetic Worlds*, in: TL Kunii and A Luciani (eds.), *Cyberworlds* (Springer, 1998) pp 19-30.

[127] NN, 'The Digital Brain'; WWW: <http://www.web-us.com/brain/digital_brain_the_extraordinary_.htm>.

[128] Ray Kurzweil is a visionary researcher. All his inventions can be found in ingenious products and successful companies. They lie in an area in which people for the time being have a firm advantage over computers – namely that of pattern recognition. Kurzweil studies how the human brains distil abstract patterns from an apparent chaos of sounds, words or images. Once these patterns are registrable, they can also be drawn off, is his expectation. WWW: <http://www.kurzweilai.net/index.html>.

[129] GE Moore, '...(T)he first microprocessor only had 22 hundred transistors. We are looking at something a million times that complex in the next generations—a billion transistors. What that gives us in the way of flexibility to design products is phenomenal.' His prediction in *Electronics*, April 19, 1965, now popularly known as Moore's Law, stated that the number of transistors on a chip doubles about every two years. WWW: <http://www.intel.com/technology/mooreslaw/index.htm>.

[130] 'Welcome to the Next Industrial Revolution'. The convergence of Bits, Neurons, Atoms and Genes. More about the context of the 'little BANG'; WWW: <http://es.epa.gov/ncer/publications/nano/pdf/RejeskiNSF(9.15.PDF> ; and WWW: <http://bang.calit2.net/>.

Technical Background of Virtual Identity

This subtitle describes the ways the virtual identity is related to the online environment. It concerns the development from anonymity to a human being with an individual Internet address, and highlights topics such as privacy, security and identity management.

Identity in Information Technology

A Virtual Identity (VID) is the representation of an identity in a virtual environment, consisting of a property of objects allowing these objects to be distinguished from each other. It can exist independent from human control and can (inter)act autonomously in an electronic system. [131]

> **Example:** When a person wants to access a computer system (or network) the identity of that person has to be converged into a unique electronic identity, to prevent confusion over names, things and people, such as a mistaken identity of one or more persons that also have access to the same system or network. To enable the interaction with the electronic system the computer software creates for that person unique objects and identities for every executing step in the communication process. When more steps in the process are connecting for end to end contact, the system keeps record of all objects and identities and links them together into one virtual ID.

Online ID

Normally the online identity – chosen by the online contributor – corresponds to the virtual identity, but in some situations it exists independently from a person. When an electronic or mechanic system needs to identify unique actions with and in the system it is necessary to have some kind of unanimity for under-standing an action that is meant to be single and unique (such as a personal message).

As mentioned earlier, the information and communication systems are generally designed to be 'identityless'. In addition, the technology that's employed operates on the principle of equality. Both are fundamental principles and each will rule the creation of virtual identities. Each time some distinction is needed the system waits for the creation of a unique virtual identity by the initiator (if that is allowed), or initiates the creation itself (such as generating a user code and password).

By default a virtual identity is automatically created – initiated by a human or a machine action – and the form is taken from a table with some kind of logic or at random. As long as the identity is not already in use in the specific situation, it can be manually chosen from a table in the system. Figure 13 gives an overview of the identity construct.

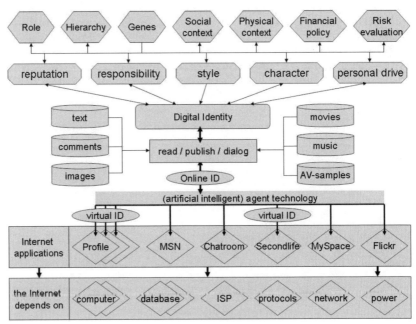

Figure 13:
Raw draft of a (con)structure of online identity © Jacob van Kokswijk 2007

---

[131] More about virtual identity in Chapter 3, page 51 ff.

Network Identification

Virtual identities are stored in a local or remote database. When computers or networks are connected, each will have its own domain. The virtual identities in one domain can be exchanged with another domain without interference, as at the moment of exchange the protocol will enhance the 'local' virtual identity by substituting some of the domain code in the header of an exchanged message (see examples of email headers 1993/2007: p 160). Furthermore, these domain codes have to be unique in order to determinate the source of a drifted virtual identity. This kind of exchange is regulated in standards and protocols, such as TCP-IP.[132]

On a technical level, all computers on the Internet use the Internet Protocol (IP) to address and to interact with each other. The two-way communication at the TCP/IP protocol levels requires that both connected parties know each other's IP address. If communications are logged (for example, by the owner of a web-based bulletin board) or intercepted, the IP address of the otherwise anonymous recipients may be discovered. Sometimes IP addresses are exposed directly as a feature of the communication system – as is often the case on wikis [133] and Internet Relay Chat (IRC) networks.[134] Casual users often do not feel that knowledge of their IP address is enough for other participants to connect their online activities to their 'real world' identities.

Depending on their technical, physical, and legal access, a determined party (such as a government prosecutor or plaintiff in a lawsuit, or a determined stalker) may be able to do so, especially if they are assisted by the records of the ISP (Internet service provider) which has assigned the IP address. Some IP addresses represent a specific computer. Others, due to proxies and Network Address Translation (NAT) may represent any number of computers or users. It is usually easy to identify which ISP assigned an address, and this may reveal some identifying information about a person, such as geographic location, or, with the use of geo-software, the affiliation with a certain organisation. To achieve strong anonymity, intermediate services may be employed to thwart attempts at identification, even by governments. These attempts use cryptography, passage through multiple legal jurisdictions, and various methods to thwart traffic analysis in order to achieve this.

> **Examples** include anonymous remailers, anonymous P2P (Peer-to-Peer) systems, and anonymous proxy services, among others. A more recent approach in Internet anonymity involves the use of an 'onion' router such as "Tor". Onion routers send information via encrypted protocols to several intermediate computers around the world in order to make identification more difficult.

The technologies as described above were surely not created or designed to facilitate these kinds of behaviours. When we look back at all incidents the technology seemed to progress too quickly, lacking a vision of the future and good planning. [135] As early as 1994 Sarah Gordon states that anonymous applications do provide a 'Use Me for Your Own Purposes' sign: [136]

> '*The anonymity of (...) these applications plays a role in the ethical models of behaviour that have developed around their uses. While FTP sites are used to transfer the sorts of programs and information with which we are concerned, there appears to be a much higher incidence of FSP sites being used on a regular basis to transfer this information and data. [137] The controversy surrounding*

---

[132] The Internet protocol suite is the set of communications protocols that implements the protocol stack on which the Internet and many commercial networks run. It is part of the TCP/IP protocol suite, which is named after two of the protocols in it: the Transmission Control Protocol (TCP) and the Internet Protocol (IP), which were also the first two networking protocols defined.

[133] A wiki is a type of Website that allows the visitors themselves to easily add, remove, and otherwise edit and change some available content, sometimes without the need for registration. This ease of interaction and operation makes a wiki an effective tool for collaborative authoring. The term wiki also can refer to the collaborative software itself (wiki engine) that facilitates the operation of such a Website, or to certain specific wiki sites, including the computer science site WikiWikiWeb, (an original wiki), and online encyclopaedias such as Wikipedia. WWW: <http://en.wikipedia.org/wiki/Wiki>.

[134] Internet Relay Chat (made by Jarkko Oikarinen) is a form of real-time Internet chat or synchronous conferencing. IRC is mainly designed for group (many-to-many) communication in discussion forums called channels, but also allows one-to-one communication and data transfers via private message.

[135] G Arnaut, 'Internet remains a work in progress.' *The Globe and Mail* (1997, March 18).

[136] S Gordon, 'Technologically Enabled Crime: Shifting Paradigms for the Year 2000' (*Elsevier's Computers and Security magazine* 1994); WWW: <http://www.research.ibm.com/antivirus/SciPapers/Gordon/Crime.html>.

[137] FTP (File Transfer Protocol, a via connection state protocol) can – in some cases – allow files to be transferred anonymously. It is a necessary thing, and its potential for abuse or misuse could be minimised by correct configuration policies.

FSP (File Server Protocol, transfers via connection list) enables having a connection only during pings, requests, et cetera. and slows down resources during inactivity.

However, the use of FSP usually requires no special privileges to set up and no special ports. It does not require separate file systems and anyone can set up this kind of `server'. This use results in the same kinds of problems with these FSP servers as with the DCC (Direct Client to Client transfer services) applications and bots (an abbreviation from robots; See note 281, p 83) that are being used to transfer viruses, virtual identities and other programs on IRC (Internet Relay Chat).

anonymity and pseudo-anonymity is one which will probably continue for a long time as we learn the effects of such freedoms. However, what we can see now is that these sorts of anonymous applications do provide almost a 'Use Me for Your Own Purposes' sign.'   Sarah Gordon

Connecting the Identity

With regard to the telephone and the Internet there is an assumption that a person-to-person contact proceeds from terminal to terminal, without taking into account the identity of, and relationship between the persons behind those terminals. Yet people make that distinction; they make it pragmatically, e.g. they differentiate between their private life and work, as well as between acquaintances and strangers. The communication technology is designed to allocate a number to each destination. Each terminal device has an address in the form of a series of hexadecimal characters. It is visible to the user as a series of numbers, sometimes translated into a letter image (URL, domain name or email address). If someone has different communication devices, they will therefore also have various communication addresses. In the Western world almost everyone has different telephone numbers for their relational environments (home, office, car, mobile) and they may have different email addresses as well.

Now, however, there is a translation model (Enum [138]) that converts a telephone number into a so-called IP address and vice versa, which brings the conversion and integration between the Internet, telephony and identity a step closer. Currently private and work telephone numbers and email addresses are intentionally kept separate and many people have several numbers and addresses. Some attempts to unite these different addresses into one personal number (such as unified messaging) have moved on and become Unified Communications. Evidently the separation of numbers relates to concealment of identity.

One of the most discussed ways of tracking and tracing an anonymous identity in a computer network is how people may 'figure out' – either spontaneously, by trial and error, or as a result of others' prompting – the various links tying into their identities that defy the effectiveness of traditional anonymity. By being alert people figure out that bar codes and numbers link to their identity when paying for purchases with a credit card. They also figure out that electronic mail – sent pseudonymously (under a fictitious name, frequently devised specifically for electronic communications) or anonymously – or Website access may nevertheless yield identifying information about them via the used computer's IP address and/or the network service provider. [139] Frequent users of the Internet realise that they become more easily identifiable through an electronic mail address that includes information about geographic location, for example, by identifying the employer or the place of access. The linkages that exist may potentially undermine the possibility of anonymity and pseudonymity and establish a correspondence between the sign under which people attempt to act and transact anonymously (or pseudonymously) and information about them.

When an anonymous person is acting on a network the links lead to the person himself. Anyone interested in watching, recording, matching, inferring and identifying, may manage to converge on individuals only with some degree of certainty, or may manage to do so by linking ultimately to that one crucial piece of information that places the unnamed person within their reach. It could be information such as the work address, the IP address, the street address, the vehicle registration, the credit card, the public transport pass, and the social security number. The chosen nicknames (or pseudonyms) of subscribers to an Internet service may serve in this way as hidden identifiers. Only when a proxy network and/or virtual identity is specially designed and developed for being anonymous, the process of tracking, tracing and identifying will take much time and will often be unsuccessful. This is one of the reasons that some opinion leaders in commerce and industry push a top-down change from anonymity to pseudonymity in order to enhance the authentication of someone's identity on the Internet. The use of pseudonyms (also known as 'nyms') enables the use of partial identities, and can also cover the entire range from anonymity to identifiability. Pseudonyms allow users to take on different identities depending on the specific context and parties involved. The use of a pseudonym is effective only when it cannot be linked to its holder or with

---

[138] ENUM unifies traditional telephony and next-generation IP networks, and provides a critical framework for mapping and processing divers network addresses. It transforms the telephone number—the most basic and commonly-used communications address—into a universal identifier that can be used across many different devices and applications (voice, fax, mobile, email, text messaging, location-based services and the Internet). WWW: <http//www.enum.org>.

[139] The location of the IP address is identified by geolocation software, such as WWW: <http://www.maxmind.com/>. This use is speedy emerging, especially for offering uncalled 'personalised' commercial and dating services. See: 'Google Seeks Patent for Targeting Ads on Wi-Fi Hotspots', 24-03-2006; WWW: < http://www.clickz.com/3593971>.

other pseudonyms a holder may have. Nonetheless, in today's state of the art technology the holder's anonymity, when necessary, the holder of the pseudonym can be revealed and as such, the person is liable and accountable for actions taken under that pseudonym.

Enjoy the Anonymity

When sending messages over the Internet, many people enjoy a sense of anonymity (or at least pseudonymity). Many popular systems, such as email, instant messaging, MSN, web forums, market places, Usenet, and Peer-to-Peer systems, foster this perception because there is often no obvious way for a casual user to connect other users with a 'real world' identity.
To address the problem of how to foster socially-desirable uses of online anonymous communication while discouraging undesirable uses, the American Association for the Advancement of Science (AAAS) held an invitational conference in November 1997. The conference was organised by Dr. Mark Frankel and Dr. Al Teich of the AAAS, and opened with 'Anonymous Communication Policies for the Internet: Results and Recommendations of the AAAS Conference' by Teich et al.[140] Teich's article discussed how anonymous communications can be shaped by the law, education, and public awareness, and highlighted the importance of involving all affected interests in policy development. The conference participants formulated some key principles, including:
(1) that online anonymous communication is morally neutral;
(2) that it should be considered a strong human and constitutional right;
(3) that online communities should be allowed to set their own policies on the use of anonymous communication; and
(4) that individuals should be informed about the extent to which their identity is disclosed online.
Another article, 'Assessing Anonymous Communication on the Internet: Policy Deliberations' by Kling et al [141] discusses the kinds of examples, issues, and arguments which animated the policy debates and which underlay the conference's more formal findings. It serves as an important briefing paper for readers who are interested in the nuances of anonymous communication over the Internet. It also serves as a tutorial about some key terms in the debates, such as anonymity, confidentiality, pseudonymity and pseudo-anonymity. It also explains some of the new technological supports for anonymous communication over the Internet.

The contradistinction of privacy and anonymity versus surveillance and information storage keeps the parties involved divided. In many countries all over the world there are more generalised concerns about privacy and information sharing.

> **UK example:** Past discussions about the proposed introduction in the UK of a National Identity Register (Identity Cards Bill, 2004; an Act in 2006). Other than the data that the UK-government itself collects, it has also passed legislation that enables the government to access data collected by other public bodies, private entities or individuals. the Regulation of Investigatory Powers Act (2000), The Terrorism Act (2000), the Anti-Terrorism, Crime & Security Act (2001) and the Regulation of Investigatory Powers (Communications Data) Order (2003) together place duties on public bodies as well as private individuals to share data under certain provisions, with high penalties for non-compliance.

McAlpine [142] pointed out that issues of identity and privacy with widespread use of e-portfolio products are rather less well explored. She highlighted the concept of identity, particularly in relation to authentication within an e-portfolio. McAlpine examined the implications and issues for awarding bodies associated with personal identity – both real and virtual – with privacy and surveillance which are raised by the widespread use of e-portfolios.

To a Safe Domain

People have different identities associated with multiple roles in specific or various contexts. These roles are generally played out within differing physical or temporal spaces, leaving the choice of how much to reveal about the other identity to the individual who inhabits it. Within cyberspace, self-presentation is to some extent controlled by the individual. However, as Suler [143] notes, aspects of

---

[140] A Teich, MS Frankel, R Kling, and Y-C Lee. 'Anonymous Communication Policies for the Internet, Results and Recommendations of the AAAS Conference'. *The Information Society*, 15(2) (1999) pp 71-77. WWW: <http://www.indiana.edu/~tisj/readers/full-text/15-2%20teich.pdf>.

[141] R Kling, Y-C Lee, A Teich, and MS Frankel, 'Assessing Anonymous Communication on the Internet: Policy Deliberations'. (1997) WWW: <http://www.indiana.edu/~tisj/readers/full-text/15-2%20kling.pdf>.

[142] M McAlpine, 'E-portfolios and Digital Identity: some issues for discussion'. *E–Learning*, Volume 2, Number 4 (2005). WWW: <http://www.wwwords.co.uk/pdf/validate.asp?j=elea&vol=2&issue=4&year=2005&article=7_McAlpine_ELEA_2_4_web>.

[143] J Suler, 'Identity Management in Cyber space' (1996) in J Suler (ed) 'The Psychology of Cyber space'. WWW: <http://www.rider.edu/~suler/psycyber/identitymanage.html>.

personality which are not consciously presented 'leak' due to the intimacy of the medium.
Suler [144] also notes that cyberspace is perceived by teenagers as a safe environment to explore issues with their identity and self-perception; however, it is noted both that there are many areas of the Web which are unsafe and unsettling, and that there is a tendency for deviant behaviour to manifest itself in cyberspace. The illusion of anonymity which it affords may encourage this, particularly in adolescents who are exploring identity issues (See: Suler and Phillips; [145] Suler [146]).

Where e-portfolios may be encapsulated as part of a safe domain – either as a segregated part of the Net, or as part of a school virtual learning environment (VLE) system – this issue may be circumvented to some extent. However, a number of systems allow and even encourage the public sharing of information, and if portfolio evidence may be used in external assessment, then by definition there must be external access to at least parts of it. Furthermore, the advantages that the ability and ease with which one can publish or share this information or share it with peers, teachers, examiners or others can bring are clear, and may well outweigh the risks. Nevertheless, McAlpine concluded, that there has to be an awareness of both the potential of accidental identity exposure and the rights of children to deem parts of their identity private and to control who has access to such information.

Online Identification

In the October 1994 issue of Wired magazine, a small news item described an obscure software product developed out of the University of Illinois. The application was called Mosaic, and it soon proved to be the "killer app" of the Internet. [147] Within a matter of several years an entire industry had been built around it and its successors. Mosaic was not the first application of its type, but it delivered a new paradigm of usability to the previously arcane task of "browsing" hypertext links. It started 'the Internet of Verbs'. Cairncross[148] states in 2001 that some of the objects connected to the Internet will not be deliberately operated by human beings at all. They will be 'thinking things', a 1999 quote of Neil Gerschenfeld, a senior academic at MIT's Media Laboratory.[149]

Figure 14

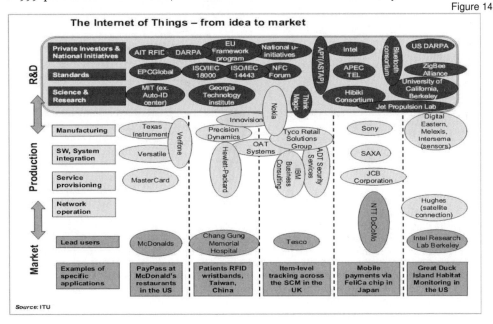

---

[144] J Suler, 'Adolescents in Cyber space' (1998) in: J Suler (ed) 'The Psychology of Cyber space'. WWW: <http://www.rider.edu/~suler/psycyber/adoles.html>.

[145] J Suler and W Phillips, 'The Bad Boys of Cyber space: deviant behaviour in multimedia chat communities', Cyber-Psychology and Behaviour, 1(2) (1998) pp. 275-294.

[146] J Suler, 'The Online Disinhibition Effect', Cyber-Psychology and Behaviour, 7(3) (2004), pp. 321-326.

[147] G Wolfe, '(The Second Phase of the) Revolution Has Begun'. Wired, Issue 2.10, Oct 1994; WWW: <http://www.wired.com/wired/archive/2.10/mosaic.html>.

[148] F Cairncross, The Death of Distance: How the Communications Revolution is Changing Our Lives, (1997; rev. 2001).

[149] N Gerschenfeld, When Things Start to Think (Henry Holt, New York 1999).

New identification technologies will enable computers to automatically recognise and identify everyday objects, and then track, trace, monitor, trigger events, and perform actions on those objects. This technology will effectively create again a new paradigm of usability, the 'Internet of things'. Today, developments are rapidly underway to take this phenomenon an important step further, by embedding short-range mobile transceivers into a wide array of additional gadgets and everyday items, thus enabling new forms of communication between people and things, and between things themselves. A new dimension has been added to the world of information and communication technologies: from anytime, anywhere connectivity for anyone, we will now have connectivity for anything. Connections will multiply and create an entirely new dynamic network of networks – an Internet of Things. In a vision of the International Telecommunication Union (ITU) [150] (figure 14) embedded intelligence in the things themselves can further enhance the power of the network by devolving information processing capabilities to the edges of the network. Finally, advances in miniaturisation and nanotechnology mean that increasingly smaller and smaller things will have the ability to interact and connect.

A combination of all these developments will create an Internet of Things that connects the world's objects in both a sensory and an intelligent manner. It contains a huge amount of humanly restrictive identification 'things' such as RFID [151], but notwithstanding the Internet of Things is also promoted by the United Nations [152].

The ITU states in their 'The Digital Life' [153] report that digital claims can be made up of sets of data, also known as attributes or identifiers.

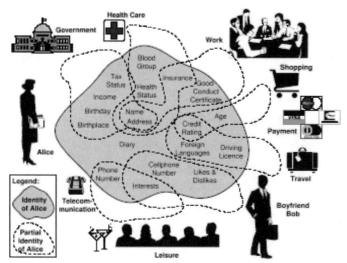

Figure 15: Identity as a subset of attributes - The many partial identities of Alice © [154]

Attributes can include a name, a date of birth, a bank balance, but also past purchasing behaviour, medical or employment records. Attributes can also include preferences, such as currency used, preferred language or seating for travel. Some information is static (such as a date of birth) and other information is dynamic (such as employer's name or dietary preferences). Attributes also ensure that the distinction between the public and private spheres of individual lives remains intact.

As figure 15 illustrates, the core of human identity is accessible only by the individual self, wherein lies the values of freedom, self-awareness and self-reflection (i.e. the 'I' of identity). The series of attributes that are accessible by external parties (i.e. the 'me' of identity) through information and communication networks must not compromise these essential values. The 'me' that is known to the outside world is a representation of characteristics that are necessary to conduct daily life within a societal and/or corporate structure. There can be many different representations of the 'me' depending on the nature of the interaction. In the information age, these representations are collectively known as 'digital identity'.

---

[150] ITU, *Internet of Things*, WWW: <http://www.itu.int/osg/spu/publications/internetofthings/>.

[151] Radio-frequency identification (RFID) is an automatic identification method, relying on storing and remotely retrieving data using devices called RFID tags or transponders. An RFID tag is an object that can be attached to or incorporated into a material, product, animal, or person for the purpose of identification using radio waves and antennas. Chip-based RFID tags contain silicon chips. Passive tags require no internal power source, whereas active tags require a power source.

[152] BBC 17-11-2005. WWW: <http//news.bbc.co.uk/1/hi/technology/4440334.stm>.

[153] ITU, *The Digital Life*. WWW: <http://www.itu.int/osg/spu/publications/digitalife/>.

[154] Source: S Clauβ and M Köhntopp, 'Identity Management and its support for multilateral security', *Computer Networks* 37 (2001) pp 205-219.

Online Identity and Privacy

Two of the distinguishing characters of the Internet are anonymity and freedom, leading to the highly experienced privacy. [155] Increasing the collection of personal data [156] results in trust destruction: the more data is collected, the more there is a lack of trust; the more security increases, the more identification... This process is becoming a continuous circle, and the loop causes an irreparable lack of trust. User's privacy can be achieved through virtual identities in Infrastructure (figure 16). There are mechanisms (such as IDsec) that provide a digital/virtual identity for users on the Internet. Users may allow Internet service providers to access their User Profile data. As such it can be an alternative to Microsoft's Passport. Emerging is the idea of "de-perimeterization" (p 108).

Online Secured Access

Virtual Identity systems provide an integrated solution to the following challenges:

- Authentication - Basically, a Virtual Identity that consists of some digital information, and that the user must be able to prove ownership of this Virtual ID.
- Authorisation - A user who has proven ownership of a Virtual Identity may be authorised to access certain information, or perform certain actions.

Web service servers must be able to verify this.

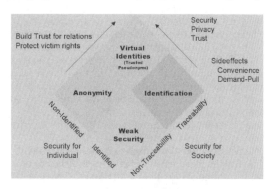

Figure 16: Identity & Trust model
Source: Open Business Innovation 2002

Prepaid ID card
A virtual identity card to protect children from paedophiles on the Internet went on sale yesterday. The device is a secure electronic identity card that displays only your first name, age, gender, and general location and is used to verify who you are chatting with online. The idea is that Net-ID-me members always swap Net-IDs with new online friends before they start chatting. This means that they know someone's age, gender, and general location before deciding whether to chat with them or not. To combat anyone getting the card, which costs £9.99, children will have to have a parent or guardian's consent, and their details must be confirmed by a professional person such as a teacher, doctor or lawyer before they can become a member. Once signed up they are then sent the card to use whenever they are online. Alex Hewitt developed the system with his daughter after he found that she had 150 people on her Instant Messenger 'buddy list' but knew the identities of fewer than 50. He said: 'The Internet is a wonderful resource and I want (children) to use it safely. This ID card removes anonymity, which is the main problem of the Internet'. Stuart Miles [157]

Identity Management Infrastructures

A problem facing anyone who hopes to build a positive online reputation is that reputations are site-specific; for example, one's reputation on eBay cannot be transferred to Slashdot. Also, scores, benefits and realised rights are site-connected.

With identity management end users are given better ways for managing their (partial) identities for specific contexts. The release of personal information can be fine-tuned to the needs of a specific transaction under the control of the end user, who also controls the links to his/her pseudonyms. When using more nicknames in online contacts some management could be necessary to keep an overview of the pseudonyms.

---

[155] Characteristics of the (original) Internet are: open structure, centre less, multimedia, communicative, identity less and free. In: J van Kokswijk, *Architectuur van een cybercultuur* (Bergboek, Zwolle 2003) p 73-4.

[156] The privacy principles for the processing of personal data are laid down, for instance in the:
Charter of Fundamental Rights of the European Union;
European convention for the protection of human rights and fundamental freedoms;
European convention for the protection of individuals with regard to the automatic processing of personal data;
Organ intelligent agency for Economic Co-operation and Development (OECD) guidelines;
EU directive on the protection of individuals with regard of the processing of personal data;
Canadian Standards Association (CSA) Model Code for the Protection of Personal Information.

[157] S Miles, 'Kids get virtual identity card for internet surfing' (2006). WWW: <http://www.pocket-lint.co.uk/news.php?newsId=4188>.

Web Address as Human ID

Multiple proposals have been made to build an identity management infrastructure into the Web protocols, even Trust models (Choi et al) [158] and "de-perimeter" designs. (Jericho Forum, page 108) All of them require an effective public key infrastructure so that the identities of separate manifestations of an online identity are probably one and the same. The future of online anonymity depends on how an identity management infrastructure is developed.

Web Address as Human ID

At the WWW2006 conference, the director Wendy Hall [159] predicted that the 'next Google' will be the first business to seize opportunities offered by rapidly developing technology. The babies of the future, for example, will have a web address instead of a National Insurance Number. [160] Hall said: 'I have a vision that in the future when a baby is born you'll get some sort of Internet-ID that is effectively your digital persona, and it will grow with you. It will actually represent you in some way - what you know, what you've done, your experiences. I guess you'd call it your URI. This is the thing that always identifies you. Every time you do something on the Internet, it is effectively logged, building up this profile that is with you for your life. Then you have your life's record, which can include any legal documents or photo-graphs or videos that you might have, that you can pass on to your children. We will be able to build software that can interpret that profile to help get the answer that you need in the context that you're in.' The idea of URI, 'the thing that always identifies you', is welcomed in the W3 scene but it has its critics too. Halpin called it 'Identity Crisis'.[161]

From a situation of minimum privacy requirements compliance, a trend towards privacy enhancing identity management is noticeable.[162] Privacy enhancing identity management combines privacy with authenticity of identity attributes. It requires technologies that allow users to control the release of personal information and to control the linkability of different occurrences of this information in different contexts. In such a scenario, generic privacy enhancing identity management solutions, if designed and deployed properly, would add value for users and could be considered as positively contributing towards privacy.

Summary Chapter 2: 'The Impact of Technology'

This chapter points out that the telegraph, telemetric, telephony, radio, television and the Internet have led to an acceleration of message traffic worldwide. Time and distance therefore acquire another dimension and new phenomena come into existence, such as 'virtual experience'. The Internet becomes an extra window from your home (or workplace) overlooking the outside world. The more often you climb through this imaginary window, the more it becomes a 'door' to the virtual society in the virtual space. The social change after the public enabling of the Internet is dramatic. Not only has the Internet caused a virtual environment that is hosting all kinds of human activities, but the Web also results in all kinds of new human behaveour. Geeks are creating your digital persona from birth to the grave.

Since the Renaissance, technology has enabled the extension of senses and imagination. People were fooled by simulation and image manipulation. Dreams and fantasies became almost real by virtual reality. Modern tele-technology has led to an enormous amount of communicating people – all over the globe – and so to an acceleration of message traffic. In the meanwhile, the Uniform Resource Identity as an individual web address for human beings should rescue the world, at least it would empower the marketing communication to a very direct online marketing. In the next two chapters the origin and development of the virtual identity are worked out. Plus the answer is given to the pressing question of 'The Lady from Dubuque'.

---

[158] H Choi, S Kruk, S Grzonkowski, K Stankiewicz, B Davis, and J Breslin, (2006) Trust models for community-aware identity management. In Proceedings of Identity, Reference, and the Web Workshop at the WWW Conference. (IRW 2006); WWW: <http://www.ibiblio.org/hhalpin/irw2006/skruk.pdf>.

[159] D Smith, 'All set for a baby.com revolution', in: *Observer* 21-05-2006 about the 15th International World Wide Web 2006 conference; WWW: <http://observer.guardian.co.uk/uk_news/story/0,,1779836,00.html>.

[160] A web address as a Uniform Resource Locator (URL) is a Uniform Resource Identity (URI) that, in addition to identifying a resource, provides a means of acting upon or obtaining a representation of the resource by describing its primary access mechanism or network 'location'.

[161] H Halpin, 'Identity, Reference, and Meaning on the Web'. Presented at the workshop on Identity, Reference, and the Web (IRW2006, Edinburgh); WWW: <http://www.ibiblio.org/hhalpin/irw2006/hhalpin.pdf>.

[162] S Clauss, A Pfitzmann, M Hansen, E. van Herreweghen 'Privacy-Enhancing Identity Management' IPTS report v 67 (2002).

3 Virtual Identity and the Internet

God has given you one face and you make yourselves another.

William Shakespeare [163]

Introduction
Throughout history human identity manipulation seems to be an obvious fact. For need or fun, people mask themselves, play with their pseudonymity, duplicate their identity with aliases, manipulate as sock puppeteers, hide behind anonymity, perform their dramatis personae, and explore their alter ago. With the ongoing development of imagination, the virtual identity is also one of the ways people extend their characters. In 2006 the announcement of registering an Internet domain for each born baby hardly upset anyone and seems to be the next step in human nature. In this chapter the various aspects of the virtual identity are explored multidisciplinary.

From Pseudo Plato to Baby.com
This subchapter points out the development from pseudonymity to cyber-entity.

The Individual
Usually you start your journey into cyberspace as an individual. In front of a computer screen, reading the characters that transform to words, you confront your singularity before building a sense of others in the electronic world. There is a double sense of individuality here. Your computer screen acts as a window to your second home, you simply connect to cyberspace by logging in, by entering your individual online name and secret personal password. Then you are rewarded with your little home in cyberspace, usually consisting of such elements as your email and your list of favourite portals and Websites. Nearly everyone spends his or her first moment in cyberspace in such individualised places. When you move from your little home to other virtual spaces this usually involves further moments of self-definition, for example, in choosing an online name for entering a chat room, in choosing a self-description at a profile site, or in outlining a biography in a dating environment. The experience of logging on occurs not only when entering cyberspace. Like passing borders and showing your passport, it is repeated as we enter name and password again and again across cyberspace.

Online Identity
Jordan [164] distinguished three key areas in which being an individual in cyberspace allows actions to be taken that are different from those in offline life. Called *identity fluidity, renovated hierarchies,* and *informational space,* these areas are briefly explored in turn. Identity fluidity is the process through which online identities are constructed. It remains true that in all sorts of online forums, an individual's offline identity cannot be known with any certainty. In the reasonably well-documented instance of 'Joan' [165] (cited by Stone as 'Julie') we are in the presence of a potential disconnection

---

[163] William Shakespeare (1564 - 1616) was an English poet and playwright widely regarded as the greatest writer of the English language, and as the pre-eminent dramatist. WWW: <http://en.wikipedia.org/wiki/William_Shakespeare>.

[164] T Jordan, 'Cyberpower. The culture and politics of cyberspace and the internet' (London, 1999); WWW: <http://www.isoc.org/inet99/proceedings/3i/3i_1.htm>.

[165] In 1985, Ms. Magazine reported on the case of 'Joan', a CompuServe user who spent a lot of time in the service's chat area. She developed close friendships with a number of other women on CompuServe. She was known for giving good advice and warm support, especially to other disabled women. In reality 'Joan' was the online presence of a conservative Jewish, teetotal, drug-fearing, low-key, sexually awkward, male, able psychiatrist, convincingly posing in the chat room as an atheistic, sexually predatory, dope-smoking, hard-drinking, flamboyant, female, disabled neuropsychologist, who lived in New York City.

L Van Gelder, 'The Strange Case of the Electronic Lover'. In: Ms. vol. 14, no. 4 (1991) pp. 199. Photocopy online: WWW: <http://www.sscnet.ucla.edu/soc/faculty/kollock/classes/cyberspace/resources/
Van%20Gelder%201991%20-%20The%20Strange%20Case%20of%20the%20Electronic%20Lover.pdf>.

AR Stone, *Will the Real Body Please Stand Up?* In: M Benedikt (ed) 'Cyberspace: First Steps' (MIT Press, London 1994) p81-118, WWW: < http://molodiez.org/net/real_body2.html>.

between online and offline identities. (Stone [166]) (Turkle)[167]

However, Jordan argues that it would be a misconception to conclude that identity disappears online. Identities that constrain, define, and categorise us exist online, but these identities are made with different resources than are used for an offline identity. Broadly speaking, online identities are constructed out of two types of indicators: identifiers and style. Neither of these mandates that someone's offline identity must reappear within their online identity, although there are many ways in which a repressed offline identity may return in the midst of online fantasy.

But which reality are we talking about when we compare the physical to the virtual? The (general) reality is something that according to philosophers (such as Spinoza and Heidegger) is fundamentally hidden in man, so that the only thing he needs to do is to develop insight into himself. Acquiring insight is a unique human ability based on the all-encompassing entity of reality. As a consequence he can work out what reality is and the more he succeeds the more balanced he becomes. Virtuality is always related to, and interacts with real, actual phenomena. Virtual identity seems to be the switch point in-between the physical and virtual realities.

Virtual Entities

What exactly is *Virtual* and what is today's *Cyberspace*?

Explanations

> **Virtual** can be imaginary, potential or substantial.
>
> The term **virtual** (Latin: *virtualis*) is a concept applied in many fields with somewhat different connotations, and also denotations. Colloquially, 'virtual' has a similar meaning to 'quasi-' or 'pseudo-' (prefixes which themselves have quite different meanings), meaning something that is almost something else, particularly when used in the adverbial form. [168]
>
> **Virtual** means: present in digital form on a monitor screen or in a computer memory. Virtual is not only conditional – you have to switch on a computer to achieve this virtuality – but it also includes something of potential, a promise of an eventual fulfilment. This virtual presence is something in itself. [169]
>
> **Cyberspace** is the space in a network of active computer memories in which a person can wander by means of his computer screen, interactively controlled by a user interface. The user stays fixed, but his/her attention flashes constantly from one side of the space to the other. Eyes (and often other senses too) follow the imaginary voyage through the virtual space. [170]

Cyber-Entities

Howard Besser [171] differentiates two entities for cyberspace: the Internet (the World Wide Web) and the Information Super Highway, each with its own content, privacy, access and infrastructure, and includes:
- Virtual societies,
- Society networks, and
- Virtual identities.

So, virtual identities are not only a part of the software environment, but also an essential part of cyberspace. As a result of the booming public use cyberspace has become a social space, with almost all characteristics of a fully-fledged world.
This so-called 'cyberworld' is based on virtual space, virtual reality and virtual society.[172]

---

[166] AR Stone, *Sex and death among the disembodied: VR, cyberspace and the nature of academic discourse*. In: SL Star, (ed) 'The Cultures of Computing' (Blackwell, Cambridge 1995) pp 243-255.

[167] S Turkle, 'Constructions And Reconstructions of Self in Virtual Reality: Playing in the MUD'. *Mind, Culture and Activity*. 1, vol. 3, (1994):158-167.

[168] Sources: Encyclopaedia Britannica Online and Wikipedia.

[169] Ref to: M van den Boomen, *Internet-ABC voor vrouwen* (Instituut voor Publiek en Politiek, Amsterdam 1995).

[170] Ref to: M van den Boomen, *Leven op het Net; sociale betekenis van virtuele gemeenschappen*; (Instituut voor Publiek en Politiek, Amsterdam 2000).

[171] H Besser, *From Internet to Information Superhighway*, In: J Brook and IA Boal (eds) Resisting the Virtual Life: The Culture and Politics of Information. (City Lights, San Francisco 1995) pp. 59-70.

Explanation:
- Virtual space is the whole of connections between elements in a massless dimension.
- Virtual reality concerns imaginations that are called up by using the virtual space. [173]
- Virtual society is the whole of participants who use those connections to socialise.

The virtual society is a result of, and exists by, the grace of tele-technology. In this society new communication behaviour with respect to technology has developed (techno-sociology), creating needs for extra or different technical possibilities, which subsequently create new expectations of 'the technology'. The different ways of communicating, gathering and spreading information, and concluding transactions in the virtual society results in different user demands on communication devices, geared towards *multiple identities,* parallel relationships and alternative (trade) transactions.

Virtual Identity

This subtitle focuses on the evolutionary relation between identity and virtual identity. It points out the human interest to mask and make their identity in becoming an 'alter ego' and in exploring their identities in an imaginary and virtual life.

Everyone an Identity

An identity is viewed as a set of self-relevant meanings – applied to the self in a social role or in a situation defining who one is – held as standards for the identity in question. (Burke; Burke and Tully)[174] Identity is related to environments and relations. (Cameron: 'Laws of Identity' page 107) [175] People may have varieties of different identities in different groups – even groups can have an identity (Surowiecki) [176] –, however Capurro et al [41] differentiate two kinds of (basic) identity: metaphysical and ontological.

But Why Virtual Identities?

The significant characteristics of the Internet are: open structure, without a centre, communicative, multimedia, free(dom) and without an identity. [177] When the Internet properties cause the loss of the regular identity, substitutes such as virtual identities will emerge and exist. As in the beginning the public Internet was an exploratory expedition into an uncultivated space that should be colonised, much of the development took place individually, without structure. Thanks to these circumstances and with the related technology features, you can equip your own world, start your own society, build your own environment, and create your own identity. Irrespective of geography, culture, race, gender, class, education, well-being, et cetera, the phenomenon virtual identity (VID) seems to have unrestricted possibilities. What exactly is Virtual Identity?

> **Virtual Identity** is the representation of an identity in a virtual environment, consisting of a property of objects allowing these objects to be distinguished from each other. It can exist independently from human control and can (inter)act autonomously in an electronic system.
>
> **Online Identity** is a persona that is implied when communicating online. It is a perceived view of who you are when online. It can exist in online connected computer systems and/or communication networks.
>
> **Digital Identity** is the representation of identity in terms of digital information.

---

[172] J van Kokswijk, *Hum@n, Telecoms & Internet as Interface to Interreality* (Bergboek Zwolle), also online: WWW: <http://www.kokswijk.nl/hum@n.pdf>.

[173] Mixed reality is the presentation of a mix of both the virtual and physical realities.

[174] PJ Burke and J Tully, 'The Measurement of Role/Identity' *Social Forces* 1977/55, pp 880-97.

[175] K Cameron, Kim Cameron's Identity Weblog. 2005. WWW: <http://www.identityblog.com/stories/2004/12/09/thelaws.html>.

[176] 'In a small group (…) the group – even if it is an ad hoc group for the sake of a single project or experiment – has an identity of its own. And the influence of the people in the group on each other's judgment is inescapable. What we'll see is that this has two consequences. On the one hand, it means small groups can make very bad decisions, because influence is more direct and immediate and small-group judgments tend to be more volatile and extreme. On the other hand, it also means that small groups have the opportunity to be more than just the sum of their parts. A successful face-to-face group is more than just collectively intelligent. It makes everyone work harder, think smarter, and reach better conclusions than they would have on their own.' In: J Surowiecki, *The Wisdom of Crowds. Why the many are smarter than the few* (Abacus 2004).

[177] J van Kokswijk, *Architectuur van een cybercultuur* (Bergboek, Zwolle 2003).

Who Am I?

> ...'Who am I?' All the rest is semantics - liberty, dignity, possession. There's only one that matters: 'Who am I?'
>
> Elisabeth, in 'The Lady from Dubuque', a 1980 play by Edward Albee

Who you are is based primarily on a description of your body and some unique registered information about the date and place of your birth, your relations and the home address. This registration gives you a formal identity that can be (re)presented by a unique official document such as a passport. This situation is totally different to the Internet, where everyone connected can acquire as many identities as he or she may want, e.g. at passport.com, and where everyone's data can be collected online, matched and processed – by companies and governments. [178]

But there is more going on between personality and identity. We can see that the younger generation (< 25 years) builds and maintains relatively confidential online relationships with members of their circle of friends, hobby club or soccer society. They read and have chat conversations which shows that tribe and group oriented (community) contacts are a replacement for the relationships one used to maintain within the family, church, bar or club. Compared by contrast with the past the zeitgeist now is more individual driven.

Within virtual communities you can position your identity in two ways. First you assign yourself 'attributes'. These attributes have to do with your choice of a name, gender, age and characteristics such as haircut, physical appearance, strength, intelligence, and so on. These attributes do not have to correspond with those of the person sitting in front of the computer. Subsequently you start 'writing' to others in that virtual room and are thus engaged in a continuous process of constructing your identity via social interaction. These virtual rooms provide a simple opportunity to play with identities (and outward appearances) and to try them out. The person who can create different identities in various rooms does not only decentralise his own personality, but can also multiply it infinitely.

After all, there is no limit to the virtual rooms, in which you can manifest yourself. In this respect electronic contact platforms and chat rooms are experimental environments in which you can discover who you are and who you want to be. Research showed that in chat rooms children succeed in manifesting several identities next to and apart from each other. [179]

Makeability

The 'makeability' of people, initially in the form of adornment using clothing, colour, and ornaments, followed in the 20th century by the use of cosmetics, all kinds of lenses, plastic surgery, orthodontics, and photomontages, has boomed due to advances in (multi) media and computer technologies.

Voices are adapted to people's wishes using Digital Sound Processing and sample technologies

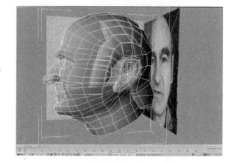

Musical compositions are turned into popular 'easy listening' with notation programs in the computer. Faces are made uniform and attractive to the public using morphing technologies. Images of people captured by means of digital cameras are fashioned into the personalities that at that time get the highest ratings.

With a computer someone can be made into an idol and with the aid of a simple image manipulation program you can transform yourself into this idol's clone in order to increase your acceptance in cyber world. (Figure 17)

Figure 17 - Electronic manipulation of an image to morph the person's face. (Screen capture)

---

[178] US Law about collecting data enables the US National Security Agency to collect the phone call records. The NSA's domestic programme is far more expansive than the White House has acknowledged before. President Bush said in 2006 he had authorised the NSA to eavesdrop - without warrants - on international calls and international emails of people suspected of having links to terrorists when one party to the communication is in the USA.

[179] K Subrahmanyam, PM Greenfield, B Tynes, 'Constructing sexuality and identity in an online teen chat *room*' *Journal of Applied Developmental Psychology*, 25 (2004) pp 651-666. WWW: <http://www.cdmc.ucla.edu/downloads/Constructing%20sexuality.pdf>.

These technologies are within reach of the general public and particularly the youngest generation is increasingly using them to 'position' their actual and/or desired image of themselves.
Now that space and time have fallen away, all traditional connections and identities are suppressed on the Internet, both the individual and the collective ones. New forms of anonymity, gender and identity switches are cultivated on the World Wide Web, thus superseding the idea of a global village. Own managed identities are becoming popular and common.[180]

Online (Mask) Identity

Wiszniewski and Coyne raise the concept of the relationship between mask and online identity.[181] They explore the philosophical implications of online identity and examine the concept of 'masking' identity. They point out that whenever an individual interacts in a social sphere they portray a mask of their identity. This behaviour is no different online and in fact becomes even more pronounced due to the decisions an online contributor must make concerning his or her online profile. The online contributor must answer specific questions about age, gender, address, username and so forth. Furthermore, as a person publishes on the web he or she adds more and more to his or her mask in the style of writing, vocabulary and topics. Though the concept is very philosophical in nature, it spurs on the thinking that online identity is a complex business and still in the process of being understood.

Figure 18: The difference and often similarity of an avatar.[182]

First of all, does the mask truly hide one's identity? The kind of mask one chooses reveals at least something of the person behind the mask. One might call this the 'metaphor' of the mask or avatar. Suler states that the online mask does not reveal the actual identity of a person.[183] However, it does reveal an example of what lies behind the mask, for instance, if a person chooses to act such as a rock star online, this metaphor reveals an interest in rock music. Even if a person chooses to hide behind a totally false identity, it says something about the fear and lack of self-esteem behind the false mask.

Second, are masks necessary for online interaction?
Because of many emotional and psychological dynamics, people can be reluctant to interact online. By evoking a mask of identity a person can create a safety net. One of the great fears of online identity is having one's identity stolen or abused. This fear keeps people from sharing who they are. Some are so fearful of identity theft or abuse that they will not even reveal information already known about them in public listings. By making the mask available, people can interact with some degree of confidence without fear.

Thirdly, do masks help with (social) education? Wiszniewski and Coyne state 'Education can be seen as the change process by which identity is realised, how one finds one's place. Education implicates the transformation of identity. Education, among other things, is a process of building up a sense of identity, generalised as a process of edification.' As pupils and students interact in an online community they must reveal something about themselves and have others respond to this contribution. In this manner, the mask is constantly being formulated in dialogue with others, and thus pupils and students will gain a richer and deeper sense of who they are. There will be a process of edification that will help students come to understand their strengths and weaknesses.[184]

Much popular music – especially the lyrics – is a ritual masking expression of one's real feeling and identity. Blogging on the Internet is also an expression of hidden personas that mask their identity with a nickname, sometimes supplied with an avatar of a fake picture. Dayal refers to 'the construction of online personas that are perhaps more glamorous or exciting than the cold facts of day-to-day reality'. [185]

---

[180] JR Suler, 'Identity Management in Cyberspace' *Journal of Applied Psychoanalytic Studies*, 4 (2002) pp 455-460.

[181] D Wiszniewski and R Coyne, *Mask and Identity: The Hermeneutics of Self-Construction in the Information Age*. In KA Renninger and W Shumar (ed.) 'Building Virtual Communities' (Cambridge Press 2002) pp. 191-214.

[182] Creative Commons general counsel Australian Lawyer Mia Garlick as avatar 'Mia Wombat' in the multi online play 'SecondLife'.

[183] JR Suler, 'The psychology of avatars and graphical space in multimedia chat communities' (2001) In: *Chat Communication* ; WWW: <http://www.rider.edu/~suler/psycyber/psyav.html>.

[184] Online identity, See: WWW: <http://en.wikipedia.org/wiki/Online_identity>.

[185] G Dayal 'Online and Under the Masks: Pop Music and Blogging Roundtable' *Education* 2005; WWW: <http//www.emplive.org /education/index.asp?categoryID=26&ccID=127&xPopConfBioID=492 &year=2005>.

Hartmann examines the multiplicities and plasticities of selves in multi-user environments, chat rooms, and other areas of textual and visual representation. [186] Focusing on a particular sub-set of textual identity expression, or typology, she contrasted the 'webgrrl' and 'cyberflaneuse'. Hartmann analysed the assumed flexibility of online identities and found that the rigid expressions of inflexibility and enculturation have emerged in practise through time.

To facilitate the online identity in communication systems a **'virtual identity'** has to be created to execute the specific online identity. It can be temporary (for the period the online contributor is online with that specific identity) or permanent (when the virtual identity stays active or mute in the online network).

The Origin of a Virtual Identity

A virtual identity is a persona that is implied when communicating online. It is a perceived view of who you are when online.
The online identity changes due to the fact that it is a visual medium with relatively low levels of truth being described. Virtual identities are the online users published personality, physical description and the ability to improvise on whoever you want to be.

Figure 19: Identity change feature (screen capture)

The online identity is one that is usually (in most cases) embellished to make the physical person be more intelligent, sexier, skinnier or bolder. The personality chosen usually embellishes what the person already has or aspires to be. It encases what the person finds attractive in the other sex, particularly in online dating and also what their ambitions are. The used 'nickname' often represents habits and characteristics of the online user, Bechar-Israeli states after research. [187] A lot of traceable plays and rituals are used in Internet Relay Chats, as Danet also finds. [188] Communication in Internet chat spaces allows participants to communicate so freely in the relative safety of anonymity that they forget their privacy. Analysing the behaviour, Scheidt concluded that adolescent females also advertise their true selves by utilising nicknames that advertise their age, actual name, self character traits, and by showing their originality by utilising innovative typography in their nicknames. [189]

Kimmelmann researched how individuals are personally extended through their virtual activities. [190] Just as handwriting is inherently an extension of the person, the portrait inevitably a meditation in the form of a painter's skill, he explored the possibilities of authenticity in electronic media.

The Shape of a Virtual Identity

Identities need to be presented and to be recognised as a unique form, otherwise a person won't recognise the specific identity. For ease of distinction and execution, a virtual identity needs to have a textual, graphical or other representation, because the technology processing in the computers

---

[186] M Hartmann, *(Un)Flexible Identities? User Metaphors in On- and Offline Discourses* (2006) in: 'Virtual Identities: The Construction of Selves in Cyberspace' by Eds. C Maun and L Corrunker (Eastern Washington University Press, Spokane 2007).

[187] H Bechar-Israeli, 'From WWW: <Bonehead> to WWW: <cLoNehEAd>: Nicknames, Play, and Identity of Internet Relay Chat' *Journal of Computer-Mediated Communication*, vol. 1, iss. 2 (1997).

[188] B Danet, Ritualised *Play, Art and Communication on Internet Relay Chat* in: Rothenbuhler and Coman, eds 'Media Anthropology' (Sage 2005) pp 229-246. WWW: <http://pluto.mscc.huji.ac.il/~msdanet/papers/ritplay.pdf>.

[189] LA Scheidt, 'Avatars and Nicknames in Adolescent Chat Spaces' (2001) WWW: <http://loisscheidt.com/working_papers_archive/Avatars_and_Nicknames.pdf>.

[190] B Kimmelman, *Retexting Experience: The Internet, Materiality, and the Self*, in: 'Virtual Identity: The Construction of Selves in Cyberspace', C Maun and L Corrunker (eds), (Eastern Washington University Press, Spokane 2007).

and networks requires it to be identified as a unique form. This essential requirement results in the naming of each to be identified object. Basically it is done by means of choosing or generating a hexadecimal code, consisting of characters such as a-z, A-Z, 0-9, and - _ . This kind of nomenclature is internationally regulated by all kinds of standards and protocols in the executing technology. In a more advanced or specialised way the presentation of the identification can also take place by means of an icon-sized graphic image or a specific sound. Essential is that all identity appearances can be recorded, stored and reproduced to interact with human beings. An unidentified alien has to be recognised in an electronic system in order to be ignored or deleted automatically.

Virtual Life

More and more people, in particular the young, start to lead a kind of 'virtual' life on the Internet through this virtual 'window'. They choose a place to stay, to show themselves, to exchange messages or to conduct transactions. They pick 'their' name, address, (mobile) telephone number, age, mother tongue, gender, looks and relationships, unhampered by their physical abilities and appearance in the regular society. A stutterer with typing skills can participate in fast discussions in a chat room without being impeded by his or her ailment. A rejected lover keeps his chin up in a chat contact and picks up full of bravado one woman after another. A lonely mother only needs to turn on her PC to get all the attention she desires after her fake picture and nickname appear at more than hundred screens worldwide. It means a change in direction in communication.

John Suler experienced the split between online and offline living and the compartmentalising of one's identities and he proposed that such behaviour is not necessarily as bad as people may think. [191] Hanging out online can be a healthy means of setting aside the stress of one's daily life. Online groups with specialised interests offer you the opportunity to focus on that particular aspect of your identity, with information and support from people that may not be available elsewhere. Dissociation can be an efficient way to manage the complexities of one's lifestyle and identity, especially when social roles are not very compatible with each other.[192]
According to Dutch studies [193] in 2006 this motive is one of main reasons for the attraction of virtual worlds such as SecondLife (the former Linden World).

> '*I often think of virtual worlds such as SecondLife as a form of escapism from the real world and all of the stresses that come with "reality." However, SecondLife is beginning to mimic real life, and a little too much so for my liking. Somehow that fine line of virtual life and real life is slowly erasing itself and the two worlds are beginning to mesh. It's interesting to hear how much real life is meshing in the virtual world and watching it all play out. As digital media becomes a way of life for us all, I wonder when the time will come when that "virtual" line is erased.*' [194]

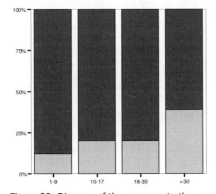
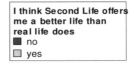

Figure 20: Diagram of the response to the question of the RL-VL relation.
The major part of the respondents says 'no'. Those who spend more than 30 hours per week in SecondLife is the largest group which says "yes" (30%).

---

[191] J Suler, 'Bringing Online and Offline Living Together; The Integration Principle' (2000); WWW: <http://www.rider.edu/~suler/psycyber/integrate.html>.

[192] See 'Sick or Side', page 31.

[193] EPN, 'The Second Life of Virtual Reality' (EPN, The Hague 2006); WWW: <http://www.epn.net/interrealiteit/EPN-REPORT-The_Second_Life_of_VR.pdf> Figures captured from the Dutch concept version.

[194] 'Second Life: A Dose of Too Much Reality'. Posted by Shannon on March 6, 2007 1:13 PM. WWW: <http://www.piercemattie.com/blogs/2007/03/second_life_a_dose_of_too_much.html>.

The moment of erasing the imagined line between real and virtual experience (that is the moment that the state of 'interreality' manifests) is not far away. The aforesaid Dutch study about SecondLife pointed out that SecondLife users still feel the difference between being in real life or in virtual life. However, they use their experience in virtual reality to increase the quality of real life, or vice versa, and their behaviour is often the same in both worlds. In real life we all use different forms of behaviour in different contexts and in different relationships. Obviously there are more constraints than apply in the virtual environment but one might consider how they relate.

Reuters opens first ever virtual news bureau in online world Second Life

New York – 16 October 2006. Reuters, the global information company, is opening the world's first virtual news bureau in Second Life. Created by Linden Lab, Second Life is an online world inhabited by hundreds of thousands of users worldwide and is one of the world's most popular virtual economies. The opening of the bureau is part of Reuters strategy to embrace new digital platforms to deliver next generation news and information.

Tom Glocer, Chief Executive Officer, Reuters, said: "Reuters is all about innovation - new technologies, new audiences, and new ways of presenting the news. In Second Life, we're making Reuters part of a new generation. We're playing an active role in this community by bringing the outside world into Second Life and vice versa."

Reuters reporter **Adam Pasick**, who will be known as Adam Reuters in Second Life, will serve as virtual bureau chief. He said: "Like any reporter, I'll cover Second Life events as they happen, interview residents and uncover interesting stories. Reuters capability and experience in news and financial reporting will be valuable to the thousands of people who need to make decisions about how they run their businesses inside Second Life. Whatever the news, Reuters will be there."

Second Life residents will be able to keep up with the latest news by using a new feature dubbed the Reuters News Center, similar to a real world mobile device. Users will be able to carry this free Heads Up Display (HUD) carrying live Reuters feeds of real-life and Second Life news wherever they go in the virtual world. In addition, residents will be able to visit the Reuters Atrium, a town hall community center, where they can meet to discuss events, see the latest images and videos of the day, or just chat with their friends. A button on their Reuters News Center will alert them to discussions and instantly teleport them to chat areas in the Reuters Atrium from wherever they are in Second Life. For Reuters SL, visit http://secondlife.reuters.com.

Figure 21: Reuters' reporter Adam Pasick as avatar 'Adam Reuters' in SecondLife

'*We had to make a few changes to our editorial practices because we're talking to people who are in essence anonymous - at least in terms of their real-life identities. So we had to get together some practices which are constantly being tested to make sure they work. Our practice is to ask people for their real-life information, now there are some people in Second Life who don't want to share that, and that's fine, but it's one of those things that I have to take into account when deciding if a source is credible. It's just one of many factors. You also look at how long they have been in Second Life, if they take an active part in businesses or organisations there or have they just created their account yesterday to be a nuisance. There are a bunch of different factors that you have to weigh together. There is no one way to do it - I think it's a pretty good analogue for the real world.'* Adam Pasick [195]

Theoretic Background of Virtual Identity

Masquerade and hidden identity are increasingly being used to examine the relationship between social, personal and bodily potential in later life. The distinction between the bodily and the digital identity and between the offline and the online identity has been expounded. Both bodily and digital identities are related to the multiple identities human beings have.

Bodily identity is conceived as a metaphysical concept when it is related to our genes, iris structure or fingerprints. These and other substantial and/or accidental characteristics, including all kinds of data about our life, hobbies, publications etc., can be digitised. With regard to identity as an ontological category, it refers to different kinds of life projects that are related but not identical. Within each medium, the body or the digital, this includes different possibilities; even, for example, the case of a permanent change of identity in a chat room. In the digital medium you may select different kinds of identities that are not identical but remain related to your bodily existence and vice versa. (Capurro)[196]

---

[195] O Luft, „Journalism', 16-2-2007; WWW <http://www.journalism.co.uk/news/story3191.shtml>.

[196] R Capurro and C Pingel, *Ethical Issues of Online Communication Research* (2000); WWW: <http://www.nyu.edu/projects/nissenbaum/ethics_cap_full.html>.

A solely metaphysical distinction between the bodily and the digital identity dissolves the richer view of existential identity. Online existence involves a bodily abstraction which implies abstraction from bodily identity and individuality. Online existence also entails abstraction from our situational orientation, an orientation which includes sharing time and space with others. And online existence is presence- as well as globally-oriented.

However, concerning presence and abstraction Naimark argues that '...A great deal of attention has been given over the past two decades to technologies for making audiovisual representations appear "more real." Photorealism in computer graphics, binaural sound, high definition television, interactivity, and (as the name implies) virtual reality are such areas. This issue of "realness" or "presence" has split the arts community into two camps: those who care and those who don't. Historical examples of this split will be discussed, including the 1900 Paris Exposition, in Abel Gance's Napoleon (1927) [197], and the Macy Conferences on Cybernetics (1947-53)'. [198]

Disembodied Information

In this age of ubiquitous networks, artificial intelligence [199] and DNA computers, information is becoming disembodied as the bodies that once carried it into virtuality. From the birth of cybernetics to artificial life, our 'back-upped' body provides an indispensable account of how we arrived in this virtual age, and of where we might go from here. N. Katherine Hayles makes the case that the body (or lack thereof) is central to this post human future. [200] She notes that the body is lost in the information age, as disembodied voices/knowledge/data came to dominate thinking about a post human evolutionary stage. Hayles also explores the development of the concept of the cyborg, and what the merger of humans and machines might eventually come to mean. She undertakes the analysis through a series of case studies. Hayles notes that the rise of artificial life will lead to the next stage of the evolution of life on Earth. 'If the name of the game is processing information,' she writes, 'it is only a matter of time until intelligent machines replace us as our evolutionary heirs.'

Most electronic security systems require some kind of identity to be pinned down; however, for many users, the charm of the Internet is precisely the ability to get away from, or play with, one's identity. That can harm the Internet trust. (O'Hara)[201]

When regular information about somebody's identity seems to escape and dissolve in space, data mining of specific body- and behaviour oriented information for risk assessment is becoming popular for the purpose of tax payment and security or safety measures. Based on characteristics such as choice of flight seating and food preferences, way of payment, countries visited, the US make profiles of travellers that will be retained for forty years.

The adage that 'Knowledge is power,' is proved daily at CBP's (Customs and Border Protection) National Targeting Centre (NTC). As the brainchild of CBP Commissioner Robert C. Bonner, the targeting centre is the centralised coordination point for all of CBPs anti-terrorism knowledge. Using sophisticated information-gathering techniques and intelligence, the NTC provides target-specific information to field offices ready to act quickly and decisively.[202]

> **American Travellers to Get Secret 'Risk Assessment' Scores** [203]
>
> **EFF Fights Huge Data-Mining Program Set for Rollout on U.S. Borders**
>
> Washington, D.C. - An invasive and unprecedented data-mining system is set to be deployed on U.S. travellers Monday, despite substantial questions about Americans' privacy. In comments sent to the Department of Homeland Security (DHS) today, the Electronic Frontier Foundation (EFF) asked the

---

[197] 'Napoleon', Silent movie about Napoleon's life (1927); WWW: <http://www.imdb.com/title/tt0018192/>.

[198] M Naimark, 'What's Wrong with this Picture? Presence and Abstraction in the Age of Cyber space'. Consciousness Reframed: Art and Consciousness in the Post-biological Era. in: R Ascott (ed) Proceedings of the 1st Intern. Conference of the Centre for Advanced Inquiry in the Interactive Arts (University of Wales, Newport 1997) WWW: <http://www.naimark.net/writing/wales.html>.

[199] S Russell and P Norvig, *Artificial Intelligence: A Modern Approach* (Prentice-Hall, 1995). WWW: <http://www.cs.berkeley.edu/~russell/aima1e/chapter02.pdf>.

[200] NK Hayles, 'How We Became Posthuman. Virtual Bodies' in Cybernetics, Literature, and Informatics (University of Chicago Press, Chicago 1999).

[201] K O'Hara, 'The ethics of cyber trust' (2006); WWW: <http://www.foresight.gov.uk/Previous_Projects/Cyber_Trust_and_Crime_Prevention/ Reports_and_Publications/Ethics_of_Cyber_Trust/kieron_ohara.pdf>.

[202] NTC; WWW: <http://www.ssrsi.org/ta1/customs.htm>.

[203] NN, American Travellers to Get Secret 'Risk Assessment' Scores (November 30, 2006: Press bulletin); WWW: <http://www.eff.org/news/>.

> agency to delay the program's rollout until it makes more details available to the public and addresses critical privacy and due process concerns.
>
> The Automated Targeting System (ATS) will create and assign 'risk assessments' to tens of millions of citizens as they enter and leave the country. Individuals will have no way to access information about their 'risk assessment' scores or to correct any false information about them. But once the assessment is made, the government will retain the information for 40 years -- as well as make it available to untold numbers of federal, state, local, and foreign agencies in addition to contractors, grantees, consultants, and others.
>
> 'The government is preparing to give millions of law-abiding citizens 'risk assessment' scores that will follow them throughout their lives,' said EFF Senior Counsel David Sobel. 'If that wasn't frightening enough, none of us will have the ability to know our own score, or to challenge it. Homeland Security needs to delay the deployment of this system and allow for an informed public debate on this dangerous proposal.'

Social Psychological Background of Virtual Identity

Every person who communicates in virtual and social environments owns a virtual identity. This virtual identity conveys an individual other than the real person and this individual is called 'virtual person' or 'virtual identity'. Although virtual identity means 'an identity that is not real', they may sometimes be identical to real identities. In fact, life is a stage on which each of us is playing out the roles that we have learnt from birth to the present day. These roles are reinforced by our culture and our education. When a person withdraws himself from the roles of a good citizen or the good friend (s)he is playing, and creates virtual identities as a result of the effects of some inner reactions, this virtual identity finds a way to express itself by existing on the Internet. Transferring feelings and thoughts to other online participants are important indicators for analysis of the virtual identity.

> **Digital Identity: Virtually You.**
> '*You are where you post. The layers of our virtual identity are complex these days: MySpace for socializing, Flickr for pictures, YouTube for movies, Netscape for news, ThisNext for products. Our blogs are for anything and everything and, increasingly, we're letting it all hang out for anyone to see. The opportunity to "just be ourselves" (or someone else entirely) on social networks presents curious challenges to our real-world identities. How do we live in both places.*' G Gould [204]

In 'The Second Self' [205] Sherry Turkle looks at the computer not as a 'tool', but as part of our social and psychological lives. She looks beyond how we use computer games and spreadsheets to explore how the computer affects our awareness of ourselves, of one another, and of our relationship with the world. 'Technology' she writes, 'catalyses changes not only in what we do but in how we think.' This treatise explores the world of virtual identity on the Internet by examining 'Multi-User Domains' (MUDs). Turkle describes MUDs as a new kind of 'virtual parlour game' and a form of online community in which one's identity (both physical and behavioural) is represented by one's own textual description of it. Turkle portrays MUDs as 'a dramatic example of how an activity on the Internet can serve as a place for the construction and reconstruction of identity'. She discussed these computer-mediated worlds and their impact on our psychological selves, describing a virtual world in which the self is multiple, fluid, and constituted in interaction with machine connections. In her concluding remarks, she points out that those MUDs are not implicated in occurrences of multiple personality disorder (MPD). Rather, manifestations of multiplicity in our culture, including MUDs and MPDs, contribute to an overall reconsideration of our traditional views of identity.

Ethical Background of Virtual Identity

Body-oriented acting in virtual communities (while living in real ones) can be considered as unethical. For example: 'Cyberwoman' is not a woman, but she is like a woman. A virtual relation with cyberwoman is the second sex in cyberspace and can destroy a real-life relationship of a

---

[204] Blog 278; WWW: <http://2007.sxsw.com/interactive/panel_picker/panels.php?mode=html>.

[205] S Turkle, *The Second Self: Computers and the Human Spirit* (Simon and Schuster, NY 1984).

S Turkle, *Life on the Screen: Identity in the Age of the Internet* (Simon and Schuster, NY 1995) Ch15.

S Turkle, 'Constructions And Reconstructions of Self in Virtual Reality: Playing in the MUD'. *Mind, Culture and Activity*. 1, vol. 3, (1994):158-167.

S Turkle, 'Cyberspace and Identity'. *Contemporary Sociology*. 28(6) (1999) pp 643-648.

S Turkle, *Parallel Lives: Working on Identity in Virtual space*. In D Grodin and TR Lindlof (Eds.), 'Constructing the Self in a Mediated World' (Sage, Thousand Oaks, CA 1996) pp. 156-175.

married couple (Westfall) [206].
The distinction between virtual and physical reality can be so very narrow (example: animated or simulated child pornography) that it can be threatening to others. Governments (US, Netherlands) forbid too realistic virtual reality child pornographic act or property (that is forbidden in the real world) however the US Supreme Court has overturned the virtual child pornography ban (the Child Pornography Prevention Act of 1996). [207]
Even so, the Internet has proven something of a boon to Saudi businesswomen, who can use the impersonality and the fluid identity of the Internet for communications that would otherwise be forbidden in face-to-face contact. (Economist)[208]
A feature of information technology is error-recovery. With a virtual identity this can be used to recover online behaviour, to reverse actions, or to achieve virtual resurrection. In the context of today's makeability it can be attractive to 'replay' the process, as if real life goodness has – to some extent – the ability to resume or restore the previous positive state of information welfare, erasing or compensating any new entropy that may have been generated by processes affecting it. (Floridi)[209]

In the non-virtual world the basic unit of reference – the individual – is one person with a single identity, a passport number, a specific address and distinct physical features. In the new information environment, the individual's bodiless existence is a username with a password and an electronic address. There is no strict correlation between the cyberian individual and non-virtual individual, as the same physical individual can appear on the Internet as several entities, each with different identification features and a different character, belonging to different communities.
The exclusive username is not even private and can be used freely by other persons.
This 'virtual freedom' introduced a revolutionary change in direction in behaviour and communications. The related technologies, such as wireless telephony, enable man to move not only his belongings, but also his contacts, and to continue his or her life somewhere else as if nothing has changed. Plug & play has become a dimension of time and space, where identity is makeable in all kinds of personas.

Integration and Separation of Identities

This subtitle focuses on the various ways of experimenting with individual identities. It shows that connecting one's online and offline lives leads to the construction of alter egos.

Offline and Online Experiments

> '...A revolution in computer interface design is changing the way we think about computers. ... The difference is that these objects also provide a link into a computer network. Doctors can examine patients while viewing superimposed medical images; children can program their own LEGO constructions; construction engineers can use ordinary paper engineering drawings to communicate with distant colleagues. Rather than immersing people in an artificially-created virtual world, the goal is to augment objects in the physical world by enhancing them with a wealth of digital information and communication capabilities.'   Mackay [210]

Identities in All Kind of Varieties: 'Who Am I This Time?' [211]
Each of us – using the Internet – has a kind of identity that is transitory as well as scattered across the web. Frequent users of social community and personal profile sites (such as myspace.com) invest a lot in creating their identity, but when a newer, cooler hangout spot comes along, they are gone, and all their messages and profile information are lost (and eventually deleted). For others, instead of 'MySpace' their blog [212] page is their virtual identity, with links to all their sub-identities

---

[206] J Westfall, 'What is cyberwoman?: The Second Sex in cyber space'. Ethics and Information Technology, Volume 2, Number 3 (2000) pp. 159-166(8).

[207] US Supreme Court Ashcroft v Free Speech Coalition, No. 00-795 (2002).

[208] NN, 'How Women Beat the Rules', *The Economist*, 30 Sept 1999.

[209] L Floridi, 'Information Ethics: On the Philosophical Foundation of Computer Ethics' (2000), WWW. <http//www.philosophyofinformation.net/ie.htm>.

[210] W Mackay, 'Augmented Reality: Linking real and virtual worlds; A new paradigm for interacting with computers'. Proceedings of ACM AVI '98, Conference on Advanced Visual Interfaces (1998).

[211] 'Who am I this time' (1982). TV Drama movie; WWW: <http://www.imdb.com/title/tt0083325/>.

[212] A blog is a Website where entries are made in journal style and displayed in a reverse chronological order. Blogs often provide commentary or news on a particular subject, such as politics, or local news; some function as more personal online diaries.

(other blogs, homepage, rss, special nicknames, friends in their blogroll). They lose all their identities unless they have the ownership of their own web domain.

Larson states that children are seamlessly switching between cellular telephones, text messaging, email, instant messaging, and Internet chat rooms to establish a constant presence in the universe. [213] They are always available, always connected, creating various (disconnected) identities for fun, attention and income. Being virtual in a virtual world is mere child's play for them.

The human behaviour in virtual societies has been observed for almost fifteen years now, and people prove to
- experiment with new forms of communication,
- explore new identities,
- create unlikely social relationships,
- search for limits in behaviour,
- construct new communities,
- execute new ways of trade, and
- enhance the borders in the physical environment.

This list seems to be confirmed by the outcome of the Dutch SecondLife study (figure 22) covered earlier, in which the motives are displayed for entering that virtual world:

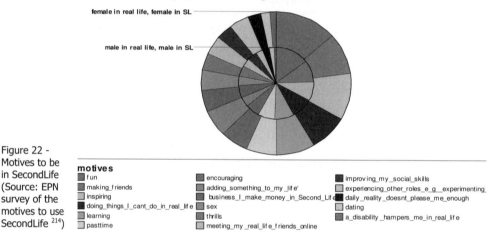

Figure 22 - Motives to be in SecondLife (Source: EPN survey of the motives to use SecondLife [214])

motives: fun, making_friends, inspiring, doing_things_I_cant_do_in_real_life, learning, pasttime, encouraging, adding_something_to_my_life, business_I_make_money_in_Second_Life, sex, thrills, meeting_my_real_life_friends_online, improving_my_social_skills, experiencing_other_roles_e_g_experimenting, daily_reality_doesnt_please_me_enough, dating, a_disability_hampers_me_in_real_life

Integrating Offline / Online

Integration can exist between an offline and an online identity. This effect is used in marketing strategies, e.g. using the online identity to make your offline fantasy real.

Suler outlines some possibilities and focuses on 'connecting one's online and offline lives': [215]

- **Telling online companions about one's offline life.**
  Lurking, imaginative role playing, and anonymous exchanges with people online can be perfectly fine activities. But if a person wants to deepen and enrich his relationship with online companions, he might consider letting them know about his in-person life: work, family, friends, home, and hobbies. Those companions will have a much better sense of who he is. They may even be able to give him some new insights into how his offline identity compares to how he presents himself online. Without even knowing it, he may have dissociated some aspect of his cyberspace self from his in-person self. Online companions can help him see that.

- **Telling offline companions about one's online life.**
  If a person lets family and friends know about her online activities, she may be allowing them to see parts of her identity that she otherwise did not fully express in-person. They can give her insightful feedback about her online lifestyle and companions. When communicating only with typed text in

---

[213] DA Larson, 'Online Dispute Resolution: Technology Takes a Place at the Table' *Negotiation Journal* (2004) pp 129/136.

[214] EPN, 'The Second Life of Virtual Reality' (EPN, The Hague 2006); WWW: <http://www.epn.net/interrealiteit/EPN-REPORT-The_Second_Life_of_VR.pdf>. Figure captured from the Dutch concept version.

[215] J Suler, 'Bringing Online and Offline Living Together; The Integration Principle' (2000); WWW: <http://www.rider.edu/~suler/psycyber/integrate.html>.

cyberspace, it's easy to misread, even distort, the personality and intentions of the people she meets. Offline friends and family - who know her well - can give her some perspective about distortions.

- **Meeting online companions in-person.**
  As friendships and romances evolve on the Internet, people eventually want to talk over the phone and meet in-person. That's usually a very natural, healthy progression. The relationship can deepen when people get to see and hear each other, when they get a chance to visit each others environment. They also get a chance to realise the misconceptions they may have developed online about each other. That, in turn, will help them understand themselves.

- **Meeting offline companions online.**
  If a person encourages family, friends, and colleagues to connect with him in cyberspace, he is opening a different channel of communication with them. Almost everyone does email nowadays, but there's also chat, message boards, interacting with Websites, online games, even imaginative role playing. He may discover something new about his companion's personality and interests. And his companion may discover something new about him.

- **Bringing online behaviour offline.**
  Cyberspace living is as 'training wheels'. On the Internet a person may be experimenting with new ways to express her/himself. She/he may be developing new behaviours and aspects of her identity. If (s)he introduces them into her/his face to face lifestyle and relationships, (s)he may better understand those behaviours and why previously (s)he was unable to develop them in the face to face world.

- **Bringing offline behaviour online.**
  Translating an aspect of one's identity from one realm to another often strengthens it. You are testing it, refining it, in a new environment. So if it's beneficial to bring online behaviours offline, then it's also beneficial to bring offline behaviours online. Cyberspace gives a person the opportunity to try out his usual face to face behaviours and methods of self expression in new situations, with new people.'

Suler states that in a process of dissociation, they don't have to own their behaviour by acknowledging it within the full context of an integrated online/offline identity. [216]

Hardey explored online and offline identities and how relationships are formed and negotiated within internet environments that offer opportunities to meet people online and move into relationships offline. [217] By analyses of user experiences of internet dating sites that are designed for those who wish to meet others in the hope of forming an intimate relationship, Hardey tried to understand the disembodied identities and interactions and the impact of cyberspace on offline sociability. Putting analysis in the context of the individualised sociability of late modernity, he argues that virtual interactions may be shaped by and grounded in the social, bodily and cultural experiences of users. It is shown that disembodied anonymity, which characterises the online experience, acts as a foundation for the building of trust and establishing real world relationships rather than the construction of fantasy selves.

Identity Construct

A virtual identity is seen as something separate; it is a fiction, a fantasy, states David.[218] When a user creates an identity, it can be a conscious construction, it can evolve subconsciously over a period of time, or it could simply be a reflection of the user in real life. Users can create an identity that is totally different from their real selves in any number of ways. They can choose gender, race, age; all the traditional delineators of real life are malleable. The identity is a pure construct of the individual and may or may not reflect reality. The ability to assume an alias, a new identity and masquerade as something that you are not, is a liberating experience. It enables you to experience life from the other side, to experience first hand prejudices and assumptions that are made by any number of social groups or otherwise.

There are two primary mediums on the Internet where virtual identities really come into play. The first is online chat (a textual medium), and the second is online gaming (a spatial medium). As games have a limited chat function, chats have a limited action function. Both serve the same purpose, to provide an alternative to the main purpose of the medium. Both functions can be used to enhance or correspond with the chat or action at hand, both can be used for comic relief, and both can help establish an identity.

---

[216] J Suler, 'The Online Disinhibition Effect', *CyberPsychology and Behavior* 7 (3) (2004) p 322.

[217] M Hardey, 'Life beyond the screen: embodiment and identity through the internet'. *Sociological Rev.* 50 (4) (2002) pp 570–585.

[218] L David, 'Virtual Identities and Quake' (1997); WWW: <http://home.alphalink.com.au/~nevyn/quakee.html>.

Tubella [219] stated (with concerns) that while traditional media, in particular television, play an enormous role in the construction of a collective identity, the Internet influences the construction of an individual identity, as individuals increasingly rely on their own resources to construct a coherent identity for themselves. This is an open process of self formation as a symbolic project through the utilisation of symbolic materials available to them. He argued that this is an open process that will change overtime as people adopt new symbolic materials. This is a relatively easy process for individuals, but much more difficult for collectivists who have the tendency to stick to their traditional values. Presenting a case about building a Catalan identity in the network society, Tubella considered what role the Internet does play in building the Catalan identity.

An example of identity construct is 'social engineering', a concept in political science that refers to (debatable) efforts to influence popular attitudes and social behaviour on a large scale, whether by governments or private groups.[220] In the political arena the counterpart of social engineering is political engineering.
Social engineering is also a collection of techniques used to manipulate people into performing actions or divulging confidential information. [221] Popular terms for these methods are 'Pretexting', 'Phishing', 'Gimmes' and (almost the same) Trojan Horse. [222] 'Quid pro Quo' is also a form of identity construct. The attacker acts as the familiar IT-helpdesk assistant and 'helps' solve a problem. In the process he obtains the user type commands that give the attacker access to and/or to launch malware.

Summary Chapter 3: 'Virtual Identity and the Internet'

This chapter discusses the factors such as communication speed, time (a moment of action), and identity that play an important role in the daily life of the average person. It underlines the diverse backgrounds of a virtual identity simply created to execute the specific online identity that is necessary in the Internet environment. As a virtual identity has to be temporary or permanent, questions arise about anonymity and traceability. The legal resolution of the anonymity issue is closely bound up with other difficult and important legal issues. Parallel social behavioural and ethical issues relate more and more to the fundamental human rights to privacy, to freedom of speech and the integrity of the own identity. More and more, the position of the identity in both physical and virtual worlds is becoming complex and causing confusion.

Who am I? Online existence involves an abstraction from bodily identity and individuality. However, different kinds of identities in the digital medium are not identical but remain related to one's bodily existence. There are always degrees of multiplicity and difference. We perform to some extent as usual – though at times we are much more aware of doing so than at other times. And yet, all these performances present who we 'really are'. We cannot separate them off from our 'true selves', even if sometimes we delude ourselves that we can. We can play with our identity, submerge our alter ego in naughtiness and paint the town red until virtual resurrection saves us. In the next chapter, the social aspects of virtual identities will be discussed.

---

[219] I Tubella, 'Television and Internet in the Construction of Identity', Project Internet Catalunya UOC2001; WWW: <http://www.cies.iscte.pt/linhas/linha2/sociedade_rede/ pr_htdocs_network/apps/immatubella.pdf>.

[220] Virtually all law and governance has the effect of changing behaviour and can be considered 'social engineering' to some extent. Prohibitions on murder, rape, suicide and littering are all policies aimed at discouraging perceived undesirable behaviours. In British and Canadian jurisprudence, changing public attitudes about behaviour is accepted as one of the key functions of laws prohibiting it. Governments also influence behaviour more subtly through incentives and disincentives built into economic policy and tax policy, for instance, and have done so for centuries. In practice, whether any specific policy is labelled as 'social engineering' is often a question of intent. The term is most often employed by the political right as an accusation against any who propose to use law, tax policy, or other kinds of state influence to change existing power relationships: for instance, between men and women, or between different ethnic groups. Political conservatives in the United States have accused their opponents of social engineering through their promotion of political correctness, insofar as it may change social attitudes by defining 'acceptable' and 'unacceptable' language or acts. WWW: <http://en.wikipedia.org/wiki/Social_engineering_(political_science)>.

[221] Social engineering is a collection of techniques used to manipulate people into performing actions or divulging confidential information. While similar to a confidence trick or simple fraud, the term typically applies to trickery for information gathering or computer system access and in most (but not all) cases the attacker never comes face-to-face with the victim. WWW: <http://en.wikipedia.org/wiki/Social_engineering_(computer_security)>.

[222] Pretexting is the act of creating and using an invented scenario to persuade a target to release information or perform an action and is usually done verbally or over the telephone.

Phishing applies to email appearing to come from a legitimate business (a bank or credit card company) requesting "verification" of information. The request contains a hyperlink to a fraudulent web page that looks legitimate (with company logos and content) and copies personal information.

Gimmes and Trojan Horse take advantage of curiosity or greed to deliver malware. It can arrive as an email attachment promising anything from a cool desktop or sexy picture to a handy system upgrade.

4 Social Aspects of Virtual Identity

'Revolutions are cruel precisely because they move too fast for those whom they strike.'

Jacob Bronowski [223]

Introduction

The social change after the public enabling of the Internet was and still is dramatic. Not only has the Internet caused a virtual environment that is hosting all kinds of human activities, it also has resulted in all kinds of new human behaviour. The ideas and opinions about virtuality and virtual subjects, objects and actions are being updated. At this moment it seems that anything can be virtual, even your identity.

> *'The space of flows is where your savings are. You think your savings are in the bank, but they are not. They are moving around electronic circuits constantly, trying to make as much money as possible. The systems are located in some financial centres such as Tokyo and Amsterdam. The space of flows are all these centres linked through computers and organised in a global network. The same is true of media and also political institutions right now. The space of flows is something that is not anywhere, and it is everywhere. Everything that matters to you in terms of the forces that shape our lives is in the space of flows, such as money, political power and media. Probably not what matters most to you: your friends, your loves, your family, your culture, your identity, your neighbours, all this is still localised in your country, in your town, in your home. But all this on which your life depends is in the space of flows. Most things that tell you who you are however are still locally based.'* Manuel Castells [224]

Signals from Society and Media

Media Technology as an Extension of Man

One hundred and fifty years after the first telegram was sent from one post office to another, we now can send messages from device to device. These are messages from person to person, from machine to machine, machine to person and person to machine. We communicate between faxes, PCs, mobile phones, etc and between different devices, e.g. fax to PDA and PC to mobile phone. Almost everyone has his or her 'own' wireless cell phone in order to communicate. A 'mobile' – because of its function, size and appearance – plays the role of a personal treasure-trove, with which we can make our way into all groups of society, across all cultures. It's a device that makes walking to and from a fixed device superfluous. An identity organiser that manages all kinds of identities and roles in electronic environments. In other words: a wireless communication device, together with the Internet has led, in just a few years time, to new forms of contact, formation of groups and the transfer of knowledge and experiences. Up to now man usually adjusted to this technology by learning how to operate the various keys and how to deal with all those weird phenomena (having to push the off key to switch something on). Each time someone replaces (the control of) an electronic device or equipment a new learning path has to be travelled to be able to use the main functions. Before we realise it, we will have a fruit basket full of devices to regulate our lives.

The Rate of Technology in the Changing Society

Technology influences changes in society and has done so throughout history. In fact, many historical eras are identified by their dominant technology: the Stone Age, Iron Age, Bronze Age, Industrial Age, and the Information Age. Technology-driven changes have been particularly evident in the past century. Electricity and steam gave us energy and power; automobiles created a more mobile, spread-out society; aircraft and improved communications led to a "smaller" world and globalisation; contraception revolutionised sexual mores; and improved water collection, sanitation, agriculture, and medicine extended our life expectancy. The reality is that the future will be different from the present

---

[223] Jacob Bronowski (1908-1974) was a polymath who found both the arts and the sciences interesting and accessible. He sought to show how these pursuits were characteristic of the identity of the human species. WWW: <http://www.drbronowski.com/>.

[224] M Castells, *The End of the Millennium, The Information Age: Economy, Society and Culture*, Vol. III. (Blackwell, 1997) p 70.

largely because of technologies now coming into existence, from nanotech Internet-based activities to genetic engineering and cloning. Society shapes technology as much as technology shapes society. The previous generation of personal digital assistants (PDAs) and palmtops, handheld computers with a screen on which you can write with a pen and sometimes equipped with a complete miniature keyboard, had to make contact with the Internet via a mobile telephone. The disadvantage of this combination (needing two hands to be able to operate both devices) was solved with the arrival of the wireless handheld computers. Such a micro-PC is now your universal remote control, but also your alarm, individual identity card, driver's license, cash card, credit card, smartcard, organiser, music player, mini-TV, library card, notebook, Internet-pc, and even more than that.

Micro-PC sticks to your body

PC's get closer to skin: Want to use Linda's mobile phone? Just speak into the bikini. Bikini phones, Tshirt Internet links and even baby-tracking nappies are close to a store near you. The American company InfoCharms believes its high-fashion wearable computers will replace the mobile phone within a few years. The company has just unleashed its range of wearable computers in fashion shows in America and Tokyo, and InfoCharms president, Mr. Alex Lightman, says wearable computers could render boxy personal computers obsolete. "These are for the chic, not for the geek," he says. "We will be making clothes that look as good or better than current standards." So, fancy a chip sewn into your son's shoe, so you can track that he doesn't stray on to the wrong side of the tracks? Or a jacket that allows you to switch on the TV, check your phone messages, tell the oven to start cooking the family meal and pinpoint your exact location--all with just a few pokes of the finger and no wires attached? Welcome to ... almost surreal world of inexpensive clothing integrated, wireless computers. Amazingly, most of these products are close to being released in the year 2000. Tim Winkler [225].

Identification possibilities for speech, handwriting and fingerprints offer the option to use such devices as a new generation of human-machine interface. Currently in an advanced stage of development is the wearable PC or "the pc you put on". This micro-PC consists of a small, high-resolution screen that, attached to a headband, is suspended on one side of the head in the user's field of vision. You can operate it with your voice and your body, as a hub, as it passes on the electronic signals, even over your skin. In the meantime fashionable clothes have been designed into which a user interface (such as a keyboard) is incorporated.

Workers can safely access the Internet in a PC-overall without needing hands, money or passport. An inner pocket holds a box with the computer and battery. A body area network (BAN) with sensors, actors and a sender/receiver ensures contact with the environment, where a personal area network (PAN) offers access to a local network (LAN) that can be connected to a wide area network (WAN). The wearable PC will in general be used to watch videos, play games and communicate via the Internet. Mobile banking, dating and gambling are other options. In the professional fields the micro-PC will be incorporated into industrial clothing for dangerous work activities (military, firemen, construction workers). The main concern regarding wireless use is the power supply. The battery capacity is still limited and prospects of improvement are not very promising.

But, in fact, the technology is so near to our body that we use it from being awake to falling asleep, and sometimes also during our sleep. What is our personality, our identity without this technology? Technology is becoming so integrated into our life and our body, as well as our environment, that it will be like an invisible part(ner) of our society.

Emotional Machine

It is said that a device cannot deal with emotions. This is not entirely correct. Exactly because of the advanced integration of wireless control and communication devices in man's daily existence, research into emotional interfaces has already been conducted. A model of product emotions based on a cognitive structure has been developed. According to this model a product reaches an emotional level when it fits in or conflicts with a personal involvement. Our involvement is more or less the stable preference for specific things in the world around us, something like our life motives.

The Frijda model (of affective computing [226]) describes three types of emotional involvement:
- goals – things we want to see happen;
- standards – how we believe things should be;
- attitudes – our characters and personal likes and dislikes.

---

[225] T Winkler, 'Wearable computers', *The Age*, Melbourne, 12-1999; WWW: <http://www.theage.com.au/>.

[226] D Moffat and NH Frijda, 'Where there's a Will there's an Agent'. In: MJ Woolridge and NR Jennings (eds): *Intelligent Agents - Proceedings of the 1994 Workshop on Agent Theories, Architectures, and Languages* (Springer 1995).

D Moffat, NH Frijda, and RH Phaf, 'Analysis of a computer model of emotions', in: A Sloman, D Hogg, G Humphreys, and A Ramsay, *Prospects for artificial intelligence.* (IOS Press, Amsterdam, 1993) pp. 219-228.

'Identity is the first thing you create in a MUD. You have to decide the name of your alternate identity, what MUDders call your character. And you have to describe who this character is, for the benefit of the other people who inhabit the same MUD. By creating your identity, you help create a world. Your character's role and the roles of the others who play with you are part of the architecture of belief that upholds for everybody in the MUD the illusion of being a wizard in a castle or a navigator aboard a starship: the roles give people new stages on which to exercise new identities, and their new identities affirm the reality of the scenario.' Howard Rheingold [227]

The next phase in PC evolution is Personalised Empathic Computing (PEC). [228] Someday, you will laugh and your computer will laugh with you. [229] When thinking of PEC, questions emerge such as: Who is the user and how to model his or her characteristics? [230] Can we control the computer's emotions, like HAL, the savvy, menacing computer in '2001' [231], whose fear that he would be unplugged led him to kill all but one of the crew members on a space mission, and how to do that? We can go further than involving emotions in the user interface.

Researcher Ray Kurzweil thinks that over 30 years he will be able to "download" his brain into a DNA-computer. But what if a downloaded mind crashes in the computer? Will life continue by means of activating a digital backup? Which identity is this backup? Kurzweil will not be held back and in 1998 launched the plan to have a chip implanted with which he could take care of the boring communication with his environment, so that, for example, the car starts when he thinks of leaving. In 'The singularity is near' Kurzweil pointed out that 'physical cloning is fundamentally different from mental cloning in which a person's entire personality, memory, skills, and history will ultimately be downloaded into a different, and most likely more powerful, thinking medium. There's no issue of philosophical identity with genetic cloning. (...) This consideration takes us back to the same questions of consciousness and identity that have been debated since Plato's dialogues.' [232]

Value and Importance of Virtual Identity for Society

The subjectivity of digital interactive communication in a social like environment is constructed by three psychosocial roots:
- networked reality,
- virtual conversation, and
- identity construction.

Riva and Mantovani state that the psychosocial dimension of interlocutor individuation has become increasingly important. [233]
In communication the 'sender' and 'receiver' – both of which are abstract, mono-functional entities – have been replaced by interlocutors endowed with thoughts, emotions, affects, and a psychosocial identity which expresses their positioning within families, groups, organisations and institutions. They also noted the increasing dematerialisation of interlocutors, or rather, the increasing irrelevance of their physical presence.

The key feature of cyberspace is the interaction through which a new sense of self and control can be constructed. The result of these new senses of self is a new sense of presence that fills the space with fluid forms of network/community. The basis of the community of people interacting in a technological environment is shifting from culture-defining mass media to a proliferation of media as alternative sources of mediated experience. We must look carefully not only at the social impact but also, and more importantly, at the technology design implications of what actually happens in networked interaction in virtual communicative environments.

---

[227] H Rheingold, *The Virtual Community*. (Harper Perennial, New York 1993); cite ref: *The Well online*: 16-05-1995.

[228] C Holden, 'The Empathic Computer', *Science* 7 October 1977: p32.

[229] D Goleman, 'Laugh and Your Computer Will Laugh With You, Someday', *New York Times* 7-1-1997; WWW: <http://query.nytimes.com/gst/ fullpage.html?res=9B06E4DD1339F934A35752C0A961958260>.

[230] EL van den Broek, 'Personalized Empathic Computing (PEC)' 2006; WWW: <http://eidetic.ai.ru.nl/egon/publications/pdf/BNAIS2006.pdf>.

[231] '2001, a Space Odyssey', 1968 movie with a humanlike computer; WWW: <http://www.imdb.com/title/tt0002022/>.

[232] R Kurzweil, *The Singularity is Near; When Humans Transcend Biology* (Viking, 2005) pp 224, 325-326.

[233] G Riva and G Mantovani, 'The need for a socio-cultural perspective in the implementation of Virtual Environments', in: *Virtual Reality*, 5, pp. 32-38, 2000.
G Riva and C Galimberti, 'Computer-mediated communication: identity and social interaction in an electronic environment' in the journal Genetic, Social and General Psychology Monographs, (1998-2000) pp 124, 434-464. WWW: <http://www.cybertherapy.info/pages/cmc.pdf> and WWW:<http://www.vepsy.com/communication/book1/1CHAPT_02.PDF>.

In his book about virtual politics Holmes focused on how virtual realities effectively extend space, time, and the body, showing how technologies such as the automobile and environments such as the movie theatre and the shopping mall prefigure cyberspace. [234] He also examines the loss of political identity in cyberspace and identifies a disembodied consumer in anonymous control of a simulated reality. The interactive networks and technologies that are said to make virtual communities possible to share with broadcasting in general, and television in particular, a reach and potential for the same kinds of imperialism of representation. Holmes' conclusion is that a virtual identity enables the rise of both Communities of Broadcast (as a mode of social integration) and Communities of Interactivity (as a mode of social participation).

Giese analyses several Mainland Chinese Bulletin Boards on issues of love, marriage, and sexuality and shows how the Internet allows its users to form and discuss opinions. [235] Presenting the case of 'Jin Yong's Inn' – a BBS for aficionados of Chinese martial arts novels – his study also shows how virtual groups replicate the process of real or offline groups, where inclusionary and exclusionary devices are practised, hierarchies formed and collective identities evolve out of long-term active participation.

The Organism is the Message

The opportunity to translate our lives, our history, our thoughts, and our skills into information enables the theoretical possibility to transmit and re-create ourselves. Wiener describes the organism as the message in his 1950 originated 'The Human Use of Human Beings'[236]: 'Let us then admit that the idea that one might conceivable travel by telegraph, in addition to travelling by train or airplane, is not intrinsically absurd, may be from realisation' in order to reduce the traffic jam because of 'the importance of traffic in the modern world from the point of view of a traffic which is overwhelmingly not so much the transmission of human bodies as the transmission of human information'. Of course the identity of the players is hidden from the observers as it is in the Turing test. [237]

The recent invention of a minuscule programmable computer made of DNA molecules in living organisms seems to substantiate his expectation that toward the end of the 21st century machines will be so intelligent that they will take over 'our' place in the evolution. Real? Allowed? The philosopher Hottois poses the following penetrating question to those who feel that modern science oversteps boundaries in its control obsession: 'In the name of which symbol, of which name with a capital letter can one subordinate techno sciences in a legitimate manner (...)? In the name of which religion, which ideology, which anthropology or which civilisation design?'[238] According to Hottois a negative answer signifies the death of the symbolising humans. It refers him to the role of supervisor of techno sciences. The only a priori (symbolic) limitation of the sciences is the potential human suffering, the resistance of (human) life: 'It may be the only ethical law that will be able to counterbalance the command of the realisation of what is possible.'

Social Paradigm Change

Using an online identity as a (second) social identity lets Internet users establish themselves in online communities and web shops. Although some people use their real names online, most Internet users prefer to identify themselves by means of pseudonyms. However, many online shopping services prescribe that you identify yourself. The result is that most online users are using role-related multiple identities. Step by step each identity is customised, including a profile with personal information, photographs et cetera. The more risk on the Internet, the more the identity changes.

Virtual dating with clever gimmicks of current cyber-fiction virtuosi is enhanced by chatting via images, such as drawings and photographs. So images are used in order to communicate about the

---

[234] D Holmes, *Virtual Politics: Identity and Community in Cyber space*. (Sage, London 1998).

[235] K Giese, Construction and Performance of Virtual Identity in the Chinese Internet, in: HK Chong and R Kluver (eds), 'Asia Encounters the Internet' (Routledge Curzon, London 2003).

[236] N Wiener, *The Human Use of Human Beings, Cybernetics and Society*, Avon Books, NY, 1967. p 129-141.

[237] AM Turing, 'Computing machinery and intelligence', *Mind*, 59 (1950) pp 433-460.

The Turing test is a proposal for a test of a machine's capability to demonstrate thought. Described by Professor Alan Turing in the 1950 paper 'Computing machinery and intelligence', it proceeds as follows: a human judge engages in a natural language conversation with two other parties, one a human and the other a machine; if the judge cannot reliably tell which is which, then the machine is said to pass the test. It is assumed that both the human and the machine try to appear human. In order to keep the test setting simple and universal (to explicitly test the linguistic capability of the machine instead of its ability to render words into audio), the conversation is usually limited to a text-only channel such as a teletype machine as Turing suggested or, more recently, IRC or IM.

[238] G Hottois, *Entre symboles et technosciences. Un itinéraire philosophique* (Seyssel, Paris 1996).

expected, desired experience. First virtually, then more or less dramatically and graphically in a telephone conversation, and finally physically. To what extent is he/she real or not real, that is the key question? Which suggestion is a teaser and what is it that must eventually become real in order to have a satisfying end of the encounter? During this entire game there is only one consideration: real or not real? Only direction at the total experience level can manage this "interreality". Because of the rapid changes in teasing expressions and pleasing techniques, a continuous weighing of 'real or not real' is required. The media turn out to be (or to have been) at the forefront regarding experience and reality and thus form part of the lie detector.

On the Internet, Everybody Can Test You're a Male Dog!

A cute cartoon dog sits in front of a computer, gazing at the monitor and typing away busily. The cartoon's caption jubilantly proclaims, 'On the Internet, nobody knows you're a dog!' [239] This image resonates with particular intensity for those members of a rapidly expanding subculture which congregates within the consensual hallucination defined as cyberspace. Users define their presence within this textual and graphical space through a variety of different activities, commercial interaction, academic research, netsurfing, real time interaction and chatting with interlocutors who are similarly "connected", but all can see the humour in this image because it illustrates so graphically a common condition of being and self definition within this space.

Users of the Internet represent themselves within it solely through the medium of keystrokes and mouse-clicks, and through this medium they can describe themselves and their physical bodies any way they like; they perform their bodies as text. On the Internet, nobody knows that you're a dog; it is possible to "computer crossdress" (Stone [240]) and to represent yourself as a different gender, age, race, etc. [241] However, the first suspense, is it a male or female 'dog', seems to be tackled very easy by the gender test. [242] Inspired by an article and a test in *The New York Times Magazine*, the Gender Genie uses a simplified version of an algorithm to predict the gender of an author.

Figure 23: At the Internet everybody can test you are a dog!

In the LambdaMOO (a text-based online virtual reality system to which multiple users are connected at the same time) it is required that one chooses a gender; though two of the choices are variations on the theme of "neuter", the choice cannot be deferred because the software programming code requires it. [243] It is impossible to receive authorisation to create a character without making this choice. Race is not only not a required choice; it is not even on the menu.

The technology of the Internet offers its participants unprecedented possibilities for communicating with each other in real time, and for controlling the conditions of their own self-representations in ways impossible in face to face interaction. The famous Peter Steiner cartoon seems to celebrate access to the Internet as a social leveller, which permits even dogs to freely express themselves in discourse with their masters, who are deceived into thinking that they are their peers, rather than their pets. The element of difference in this cartoon the difference between species, is comically subverted in this image; in the medium of cyberspace, distinctions and imbalances in power between beings who present themselves solely through writing seem to have been deferred, if not effaced. The accelerating technology and progressive media influence seem to erode our sense of who we are.

---

[239] Cartoon by Peter Steiner (*The New Yorker*, NY: July 5, 1993) Vol. 69 (LXIX) no. 20, page 61. WWW: <http://www1.umw.edu/~ernie/comic2.gif>.

[240] AR Stone, Will *the Real Body Please Stand Up?: Boundary Stories About Virtual Cultures*. In: B Michael (Ed.) 'Cyberspace: First Steps' (MIT Press, Cambridge, 1994) p84.

[241] 'I loved your email but I thought...' WWW: <http://www1.umw.edu/~ernie/comic1.gif>.

[242] Gender test, developed by M Koppel (Bar-Ilan University in Israel) and S Argamon (Illinois Institute of Technology); WWW: <http://bookblog.net/gender/genie.php>.

[243] L Nakamura, 'Race In/For Cyberspace: Identity Tourism and Racial Passing on the Internet'; WWW: <http://www.humanities.uci.edu/mposter/syllabi/readings/nakamura.html>.

Societies as well as individuals construct identities through the expressions and actions of people, the physical being of humans, the structures of our cities and institutions, and the cultural record we leave by design or by happenstance. As a society, we are the total sum of our multiple meanings. As individuals, we have, throughout the long course of human history, found our identity in the relationships we have with those close to us in our families and communities. After opening the traditional local environment to a global scene, we must include the Internet as a space in which we create our identities.

Molony considers 'what it is that constitutes our identity? [244] Is it our selves, interacting with one another across cyberspace? Have our identities become modified or invented versions of our physical selves? Is our identity as a culture a vast morass of unmediated information? Is our identity as a society eroding because we are unclear about the survival of our cultural legacy? What is the potential of the Internet to build or disrupt the communities that underpin our identities? Do cyberspace connections bring us together or segregate us; do they make us more or less lonely?' However, we can be optimistic about our ability to craft new identities through the Internet that can, paradoxically, help us forge old-fashioned connectedness in a seemingly disconnected, modernising world.

Fantasy, Media, and Identity

In this section the influence of the media on the development of fantasy and identity is worked out. Media meet the need for both truth and fantasy. By using the media to experiment with identity, people satisfy themselves in the social environment.

The Media Meet the Need for Truth

Popular news sources on the Internet (such as the bloggers and e-zines) engage in 'framing' by selecting, emphasising, omitting and detailing news. The reader's organised thoughts gives a context/deeper meaning to news facts, in particular when a news item answers four related questions:
1. What is the problem?
2. What caused it?
3. Who/what is guilty/morally responsible?
4. What is the solution/remedy and who can offer it?

The supposed degree of truth of these news sources is as high as the reader's media opinion, combined with the own (private) opinion, leads to a public (publicly expressed) opinion, after which it also becomes the reader's private opinion. That opinion influences the actual role and identity of the reader, not only in real world, but also in the online world.
For the visual media on the Internet (such as webcast and YouTube) the same influencing dangers threaten to occur as for television: when you watch too much television and the offered programmes are monotonous, the viewer's reality image will become identical to the reality image in the television programmes offered, irrespective of the degree of reality of such programmes. The scope of this book does not cover the question what the value of the ordinary media is in the viewer's experience and how it would develop if interactive television would actually be available to everyone. However, some predictions can be made. In the discussion about the relationship between television and violence, empirical research has confirmed that imitation in identity and behaviour (Ref: modelling theory of Bandura [245]) and disinhibition (disinhibition theory [246]) play a role.

---

[244] B Molony, 'Is rapidly accelerating technology eroding our sense of who we are?' Report of a panel meeting. (2001); WWW: <http://www.scu.edu/sts/nexus/summer2001/MolonyArticle.cfm>.

[245] A Bandura, *Social Learning Theory*. (General Learning Press, New York 1971/1976/1997).
'Learning would be exceedingly laborious, not to mention hazardous, if people had to rely solely on the effects of their own actions to inform them what to do. Fortunately, most human behavior is learned observationally through modelling: from observing others one forms an idea of how new behaviors are performed, and on later occasions this coded information serves as a guide for action. (...) Social learning theory explains human behavior in terms of continuous reciprocal interaction between cognitive, behavioral, an environmental influences. The component processes underlying observational learning are: (1) Attention, including modelled events (distinctiveness, affective valence, complexity, prevalence, functional value) and observer characteristics (sensory capacities, arousal level, perceptual set, past reinforcement), (2) Retention, including symbolic coding, cognitive organization, symbolic rehearsal, motor rehearsal), (3) Motor Reproduction, including physical capabilities, self-observation of reproduction, accuracy of feedback, and (4) Motivation, including external, vicarious and self reinforcement.'

[246] R Gustafson, 'Alcohol and aggression'. *Journal of Offender Rehabilitation* 21(3/4):41–80, 1994.

Watching television for hours has measurable effects on the viewers' experiences. [247] Additional empirical research has shown that the longer you watch, the more serious you experience a problem. [248] For example, someone who watches four hours of television takes in 9% more crime than someone who watches two hours.
On the other hand, many frequent-viewers considered the world to be a frightening place. [249] Does watching television or surfing for long periods of time automatically lead to criminal behaviour? Or to neurotic, frightened people? Or is this the difference between young and old? Anyhow, the younger generation is actively looking for 'clarity about reality' and for the link between those physical and virtual realities, the total experience. Whether they do so by bungee jumping or by channel surfing does not make much difference with respect to the end goal. A Virtual Identity is part of their (and your) Personal Universe.

Figure 24. [250]

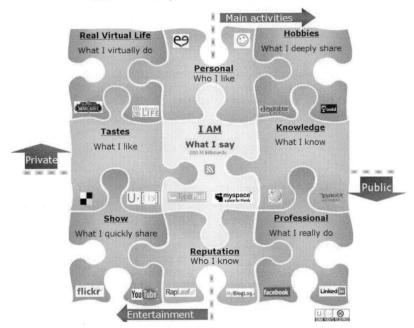

The Media Meet a Need for Experience

As mentioned before, not only the message *via* but also the message *of* the medium plays a role. When handwriting was invented the human senses (sight, smell, feeling, hearing, touch and taste) were in a natural balance. Printing technology disturbed this balance and the one-way face gained the upper hand: intellect first, feeling second. Feeling becomes linear, regular, repetitive and logical.

The invention of the television was followed by a return of the multi-way face and hearing. Music styles emerge and are intensified after the invention of the new media. Everyone sees and hears everything, including what they are not intended to hear and is often meant for different situations. You receive information aimed at a different gender, age, family situation of culture. This brings along yet again new culinary, clothing and fighting styles. The media's benefit in meeting human needs becomes clear in the so-called Uses & Gratifications model. [251] It shows that the media do not influence people, but that people have motives to use the media.

---

[247] MM Lefkowitz, 'Television violence & Social behavior, a follow-up study over ten years'; in: GA Comstock and EA Rubenstein (eds) *Television and social behavior* (1972).

[248] G Gerbner, *Toward "Cultural Indicators": The Analysis of Mass Mediated Public Message Systems* (1969).

[249] WJ Severin and JW Tankard, *Communications Theories: Origins, Methods, and Uses in the Mass Media.* (Longman Publishers, New York 1997) pp 199-252.

[250] Virtual Identity ~ Personal Universe. Source: WWW: < http://ulik.typepad.com/leafar/2006/10/ulik_unleash_id.html>.

[251] JG Blumler and E Katz (eds.) 'The Uses of Mass Communications: Current Perspectives on Gratifications Research' (1974), WWW: <http://en.wikibooks.org/wiki/Communication_Theory/Uses_and_Gratifications>.

Such motives are:
- entertainment;
- personal relationships;
- personality (and identity) development;
- surveillance;
- information in order to orientate oneself;
- the need to act in a social/economic/political environment.

Media cannot generate the effects anticipated by the individual or the conclusions that are drawn. People do those themselves. The individual mood is compensated by the choice of television programme: when stressed the viewer watches more television, in particular relaxing programmes, games and shows; when bored the viewer zaps to violent programmes.

Thanks to the remote control the television had become an interactive mood gauge: real reality is far away. What's real? Integrated music and imagery play this role in a youth's life. Born in cyberworld with a mobile in the crib, a playstation as teddy, the Internet as mother's milk and TMF as wallpaper, the new media create their identity and environment. A chat acts as their sounding board. Create yourself an image, an identity, make yourself beautiful! While most would call 'The Matrix' series [252] the ultimate film about virtual reality and man's fight to overthrow the machine empire, there are a couple of other films that share the Matrix's theme of virtual and real worlds blending together. "eXistenZ" is one of the better low-budget movies about a game designer on the run within the world's most advanced virtual reality game. [253]

Writer and director David Cronenberg spins an unusual tale that is an interesting watch for fans of 'The Matrix'. A similar story is 'The Thirteenth Floor', a film that blends together reality and a virtual world creating an intriguing story about what is real and what isn't. [254] It is a movie that has much to do with virtual reality.

Fantasising Meets Human Needs

Fantasising and exploring reality are human qualities. Fantasy worlds have existed from time immemorial in games, from dice to doll's houses, and from court poetry to theatre. As mechanisation and automation advanced, mechanised games were brought on the market. Fantasy appears more real than the real world itself.

Playing in a fantasy world has its limitations, however graphical it may be. There comes a moment in which fantasy and reality touch each other and the fantasy will take place somewhere in the physical world. Such a fusion of the visual environment with the virtual world results in a new experience. The related technology has already been available for quite a while. In the past few years various companies have photographed the world in great detail and stored this in a database. Navigation systems, cartographers, appraisers and inspectors can now make use of it. Google Earth shows your location in a real world's aerial view, including street names and traffic directions. The Earthviewer3D and advanced flight and space simulators combine the images from an enormous digital atlas of the world, consisting of countless satellite and aerial photographs, creating the effect of flying across actual land. Motion can be captured in pictures. The files and software are also available to the general public. So people who use the Internet (and have a broadband connection) can fly themselves from, for example, Amsterdam to Bagdad. Each flight can end in gruesomely detailed war pictures. Using this technology, combined with wireless telephony and location-based information, it is then possible to position yourself in a 'real' environmental image, enabling you to see where you are.

In the 17th century three-dimensional decors were made for comedies to help the fantasy really experience the plot. The current reality technology enables people to lead an almost real life in a world that is not physically tangible, to usually do what they also can do in the daily world around us. Travelling, visiting museums, leafing through stacks of library books or getting a flying lesson. We can even start to love someone, just for what and how he or she types something or for the photo or drawing that is shown. We can marry him/her, die or feel ill. The only difference between cyberworld and the physical world is the existence of your body. As long as time-travel has not become reality we cannot virtually move our bodies. We feel the pain of the mouse movements in

---

[252] The Matrix trilogy are movies by the Wachowski Bros. (1999-2003); WWW: <http://www.imdb.com/title/tt0133093/>; WWW: <http://www.imdb.com/title/tt0234215/>, and: WWW: <http://www.imdb.com/title/tt0242653/>.

[253] Movie "eXistenZ" is about a virtual-reality game that taps into the players' minds. WWW: <http://www.imdb.com/title/tt0120907/>.

[254] Movie 'The Thriteenth Floot' is about reality. You can go there even though it doesn't exist. WWW: <http://www.imdb.com/title/tt0139809/>.

our wrists and our backs are aching from sitting too long behind the keyboard. Furthermore there is a palette of accessories to connect our body to the computer by which (y)our senses can be extended. How more senses can be remotely triggered, how more imagination can be powered.

Categorisation of Fantasy Needs

The foregoing shows that people have a need for fantasy, both in their dreams and the physical reality. Their fantasy has been extra stimulated by the arrival of virtual reality. Step by step the technology brings us more intense levels of mixed reality. The seamless stay in both real and virtual reality, the so called "interreality" is happening. If you can wake up reassured from an exciting dream, then you literally go from one reality to another.

The need arises to keep an 'experience overview' in order to classify each form of experience. The need for determining and categorising experiences in the human "Interreality" fits in well with the studies and theories of McLuhan and Hunter.

McLuhan endorses this development with a casuistry, in which his students who read comic books have a better starting point to understand the media than he does. [255] Comics appear to involve the reader into the story in a more imaginative way and make a useful contribution to understanding the positioning in, and manipulation of time and space.

> **WebCamLife** Why is somebody publishing almost all private information and experiences on the Internet?
>
> In 1996 the 20 year old Jennifer Kaye Ringley started a popular Website whose main feature was several webcams that allowed Internet users to observe the life of a young woman. The site was online from April 1996 until the end of 2003. [256]
>
> In 2002 the 24 year old Kim is a 6th year medicine student in The Netherlands. The forthcoming neurosurgeon has her own site where she presents herself daily with a webcam and a dairy. Once her friends had made their own homepages, Kim wanted to do so too. With her boyfriend's webcam Kim subscribed to some webcam classification sites. Her virtual window has become popular worldwide with a score of about 1.500 hits per day. One-third of the site visitors are recognised from the USA, and (only) one-eight part from The Netherlands. Kim likes most of the reactions but notes: 'I don't like that my colleagues discover my site, because they will get a wrong image of me. Not because I am on-line different from who I am in real world, but people could imagine so. My webcam and photographs have nothing to do with sex, but people who don't know me so well, take their narrow conclusions, and I regret that.'

Since as early as 1948 human media needs have been macrosociologically categorised. (Lasswell et al [257]) Referring to empirical research, in which 35 needs were accumulated which could then be divided into five categories, all media users appear to have (in essence) the same needs:

- Cognitive Needs – Needs related to strengthening of information knowledge, and understanding of our environment. Acquiring information, knowledge and understanding.
- Personal Integrative Needs – Needs related to strengthening of credibility, confidence, stability, and status of the individual.
- Social Integrative Needs – Needs related to strengthening of contact with family, friends, and the world.
- Affective Needs – Needs related to strengthening of aesthetic, pleasurable, and emotional experiences.
- Escapist Needs/Tension release needs – Needs related to escape, tension release, and desire for diversion.

Christopher Hunter has worked out this model for the World Wide Web and identifies the social and psychological needs of the Internet user. [258] In addition he has explored whether the Web offers the content to satisfy these needs. The Internet users' needs on the Web turn out to be no different from the other media sources and communication channels. In this context Hunter considers the World Wide Web – in line with McLuhan – as both the messenger and the message.

---

[255] M McLuhan, *Understanding media, The extensions of man*. (MIT Press, Cambridge MA 1964/1994 rev.).

[256] See more: WWW: < http://en.wikipedia.org/wiki/JenniCam>.

[257] HD Lasswell, D Lerner and ID Pool: *The comparative study of symbols* (Stanford University Press 1952) p12.

[258] C Hunter, 'The Uses & Gratifications of the World Wide Web' (1996); WWW:
<http://www.asc.upenn.edu/usr/chunter/webuses.html>.

Real Imagination in Virtual Affairs

Business as Booster

Cyberworld is not a truly complete world as you cannot eat a snack, drink a beer, or become ill in it. Also in the area of communication it cannot become a life such as reality unless man adapts genetically in the same way as nature evolves when the environment physically changes. The way in which the youngest generation (born in the digital age, after 1980) familiarises itself with this communication technology creates expectations as a matter of course. Young people keep their multiple identities 'in stock' in order to explore the digital playground from different views. Thanks to the rapid technical development, the quality of the fantasy role-playing games has increased to such an extent that they have the character of a simulator for virtual reality.

Identity, Time and Origin

The meaning of identity is changing, as identity is now defined at the place and plot you are in. It is defined too by relationships such as membership of a wider grouping.

Factors such as communication speed, template (as moment of action), identity and the short service life of products and services play an important role in the daily life of the average person. They are even more important for the generation that has grown up with them and therefore accept the speed of communication, firm reaction deadlines, and trends and hypes that influence their identities. The Internet is another speedy step, because now the relations switch from what is happening on the Internet to what happening in real world.

I have noticed that the young (in order to come to a choice/decision/result) set *deadlines* for themselves in the era of messaging. Time is the message, not the message itself. They want to know/have something/someone '**now!**'. Often they send a question into cyberspace, shooting pellets at everything that moves and each reaction is sniffed at and checked out for value/usefulness up until a certain moment; then it is no longer useful or even becomes undesired. From that moment on they start to ignore them. In certain situations they even physically close off reception, e.g. by logging out when they are online (to subsequently log-in again under a different nickname [= pseudo-identity]) or by using a special email address or mobile phone number they no longer use after the deadline (after some months they automatically die a virtual death by automatic reset). The younger generation uses 'ignoring' to rule their world and they can do so because of the pseudo-anonymity of the Internet.

Social View to Internet

Just as voice is still the 'killer application' for fixed and mobile telecom networks, the Internet is all about contacts. Except for some negative effects, the main image is positive, e.g. dating, consultancy, distance learning. Fewer senses are used by the graphically-based Internet contact, enabling more openness in the contact. Surveys on medical consultancy, and distance learning indicate that people are more 'open' to machines than to real persons. 'I know my students better by distance learning than in real contact.' [259] Being hidden (invisible) seems no barrier to more intimate contact.

Time in Virtual Space

The core of the current "anytime, anywhere" electronic communications is that it is an actual presentation of yourself in the virtual world. This subchapter concerns the factor time in our offline and online existence. Time is the message!

Space is pointed as the infinite extension of the three-dimensional region in which all matter exists, but is related in real world to elements or points satisfying specified geometric postulates. A message relocates from one point to another. Whereas in the virtuality space is infinite (and therefore unlimited), time is no longer an isolated (from the message) fact. Because of the speed with which nowadays messages can be sent and the many ways in which the moment of arrival can be steered/manipulated, from real time to low priority, messages are hardly '"timeless" anymore. The monitored chats and some interviews show that users of electronic means of communication, when receiving a message, not only consider the content of the message and the medium used to send the message (McLuhan's 'the medium is the message' theory) to be part of the message, but the time of perusal of the announcement, arrival and/or essence of the message as well.

---

[259] Spoken quote of Professor Ian Lloyd, University of Strathclyde's Law School, 04-12-2006.

Time is a Factor

Time is indeed linked to messages and even forms a message in itself. The short duration between sending and receiving is of decisive importance in many contact and transaction moments in relationships and company processes. It applies to instant messages as well as delayed and prioritised messages. The fact that time can be seen as a message is an extension of McLuhan's theory that the medium is the message.

In order to gain insight into the importance of 'time' as a message it is necessary to look at the way in which people mutually make contact (and enter into relationships). This approach has been changed in a time span of one third of a century. People once used to enter only into relationships within the family, church or club (under penalty of isolation in the community). So they had to stay in that community. Nowadays you can be wherever you want to be and have multiple parallel personal contacts that no longer as a matter of course are based on physical contact.

The telephone and the Internet enable people to cross the boundary of contact in milliseconds; such contacts appear to last even on an anonymous basis. The basis of each relationship is formed by the common interests and via each connection (usually interest-oriented) information is exchanged. It no longer matters who you are, but what you are, know, are able to do or which relationships you have. Collecting (contact addresses of) 'friends' via electronic chat groups, news lists and discussion platforms increases your possibilities to exchange information. The longer your contact list, the faster you can make contact, the better you are informed, the more support you can ask. Thus the ways to acquire knowledge and skills, share findings, exchange experiences and conducting transactions have been drastically changed. On the basis of, among others, common interests and often gender multiple, star-shaped and parallel relationships with groups/group members are formed, both in daily life and the virtual world (communities of interest/tribes). The start pages (portals), the search engines (Google, AltaVista), and the social sites (You Tube, My Space, Face book, Lycos) point the surfer to the centres of subject-linked groups.

> '**Welcome to the Tiger at Home**. A Website about long fingernails on guys and men. It is about the positive aspects of long nails and deals with issues like the "why" and "how" of long nails, scratching and pinching, the social issues etc. with a great collection of pictures contributed by a lot of people, video-clips, a message board facility to share information and opinions and room for (fantasy or real-life) stories.' [260]

All over the world contacts are made about specific subjects, based on anonymity and equality, and usually also on open-mindedness. When contact has been established (and the technical contact information has been added to the contact list) the contact is maintained for as long as there is time and interest. Interest is usually linked to topicality, and therefore to time. Sometimes the members of groups meet, usually at large-scale theme meetings somewhere in a country. Sometimes a virtual contact leads to a usually once-only and sometimes frequent meeting in the flesh, as monitored chat reports show. "Seeing" each other is absolutely not required to have a good contact. In many cases it even backfires.

Being Up 2 Date

Actually, a separate chapter could be devoted to real time communication via media such as the Internet and mobile telephony (in particular the text forms, such as Internet Relay Chat, chat and sms). The core of the current "anytime, anywhere" electronic communications is that it is an actual presentation of yourself in the virtual world, sometimes one-on-one (monocast), often one-on group (select cast) and sometimes one-on-all (broadcast).

You can express yourself in letters, colours and images. Your passport photo 'speak' in Lycos' Club Eden. It all happens online and in real-time, in which a message lapsed in time usually is no longer a message: after all, a message is sent at a specific time and old news is ignored. New and news lie closely together. News is no longer news if someone else knows it already. So why waste time to read something that is 'dated'? Even Websites are checked for topicality... Does the offered news seem dated? Then do not waste time reading it! The ease with which internet users ignore their language errors has everything to do with time: correction takes time, and if you are not careful, you will lag behind! I Am, Therefore I Chat.

Your chosen identity sometimes chats with one person, but usually with several people at the same time. Whoever follows a chat as a participant discovers upon entering the chat room a restless screen of lines of text that go back and forth. Via the scroll the lines go by. The sentences appear on

---

[260] Tiger at home: Liking Long Fingernails on Man. WWW: < http://www.xs4all.nl/~richardw/nails.htm>.

screen in order of receipt at the server and new sentences push away the oldest ones. To an outsider the mass of text will seem a hotchpotch of words that hardly seem interconnected.

Microsoft's latest Messenger version combines your contacts, scheduler and messages with a text/speech interface. The interface to a wireless telephone and a video interface based on the build-in webcam completes the integration of contacts. These kind of integrated functions are already build into games such as 'World of Warcraft' and virtual worlds such as SecondLife.

> **Visual chat** *is a simple way to describe them such as multimedia chat, GMUKS (graphical multi-user conversations), and 'habitats', (...) They are something of a cross between a MOO and a traditional chat room. As social environments, they are unique in that they are graphical. Rather than limiting users to text-only communications, as in most chat rooms, multimedia programs add a visual dimension that creates the illusion of movement, space, and physicality. It allows people to express their identity VISUALLY, rather than just through written words. The result is a whole new realm for self-expression and social interaction with subtleties and complexities not seen in text-only chat rooms (...) 'avatars' or 'props'. Although these words often are used interchangeably, there is a slight distinction in the minds of some users. Avatars refer to pictures, drawings, or icons that users choose to represent themselves. Props are objects that users may add to their avatars (say, a hat or cigar) or place into the Palace room or give to another person (say, a glass of beer or a bouquet of flowers).*
>
> John Suler [261]

Today's communications are action-oriented and custom-made for individuals and/or target groups that employ various channels (web, mail, television, sms). The contact is very interactive but also superficial, with simplified language and relations. Much-used terms and abbreviations are included in new editions of dictionaries. The finite form in communications is the same. Hierarchy has deteriorated into urgency or counts in exchange for a business deal. Gender as a distinguishing factor is no longer the leading theme. On the Internet you are equal and yet different, the latter being more the result of one's own choice than of origin. Etiquette on the basis of standards and values is converted into regional self-regulation from a practical point of view: ensure that you are on time and up to date.

The new telecommunications technology has lowered barriers in our association with each other. Manners are simple, everyone is 'you', there is no etiquette and little hierarchy. It is unisex all the way: differences between men and women are – when essential – reduced to basic gender differences. In summary you can say that communication via the Internet and wireless telephony is quick, action oriented, interactive and intensive. Finite forms are (all) reduced. Transfer takes place simultaneously through several media and sometimes at the same time (multi-channel). Contacts are maintained generally and very personally. In this hectic 21st century communication with people who can be reached at various places during the day also includes making appointments and keeping them, planning, taking responsibility, making choices, evaluating, etc. Qualities in which many older adults do not excel, but which are taken for granted by the youngest generation. They manage their time and actions, related to their identities.

Time has Many Faces

What is time? This paragraph discusses linear time, time as experience and time as a message. Speed in interaction stands for young and dynamic. Young women, the grrrlz, call a man over 50 a 'fossil' because he lacks quickness of response. It makes clear that thinking in more scenarios at the same time and carrying out all kinds of parallel sending actions do not only make higher demands on technology, but also on the 'receiver'. When someone does all kinds of different things at the same time, using multiple devices (so-called interfaces) and various communication channels, the physical factor 'time' will become important – next to the mental 'stress'. Communication channels and means of presentation have speed reduction qualities in the sending process and the human brain cannot process everything at the same speed. This is why, for example, notifications (so-called alerts) and messages do not arrive synchronously to the original moment of sending. It not only leads to awkward misunderstandings but also to erroneous (trans)actions. Interactive games expect a quick reaction (for example: a shooting game as 'BZFlag' requires a quick response, otherwise you will 'fire' at the place where your opponent was, while he has already moved into another position, returns fire and hits you). Second item is that in many web and chat sites everyone else can take over the identity of someone who just quitted the site. On the Internet nobody knows you're not the same dog. Synchronism and response time will play a key role, same as authenticity. It has consequences for our society but will also make heavier and different demands on the front offices and technical systems of companies and authorities.

---

[261] J Suler, 'The Psychology of Avatars and Graphical Space', 1996/2006 rev. WWW: <http://www.rider.edu/~suler/psycyber/>.

Time is money, the old economy says. Almost true, in any case it is a loss of time to correct your mistakes when they do not really matter. Whoever corrects the mistakes in his communication lags behind and loses the other party/parties. Waiting for someone who is making useless improvements is a waste of time. It devalues your value in the community. So you sever the contact and focus on another candidate. When you return after an absence you will peruse the most recent newspapers, open the most recently dated e-mail messages. You do not bother with the rest and in all the hustle and bustle these messages will disappear from the recipient's sight (and attention), often forever. In the new economy everything happens in real interaction, more quickly and on time, as the parties involved expect a direct response. Your virtual identity is temporarily. The saying 'out of sight, out of mind' has turned into 'out of date, out of sight'. Just mind the "realitime"…

Time and Identity

Factors such as communication speed, time (as moment of action), identity and short service life of products and services play an important role in the daily life of the average person. They are even more important – or better: self-evident – for the generation that has grown up and therefore accept the speed of communication, firm reaction deadlines, and trends and hypes that influence their identities. We are familiar with the message (content) and the medium as the message (see: McLuhan), but when we ponder what is going on in the virtual world, there is also the time of receipt/taking note of the message (messenger).

> **For example**: a telephone call at night (when already asleep) carries a different message than one during daytime. When after a message on the school board that pupils who failed the final exams will be called between 4 and 5 p.m., the telephone rings at 4.23 p.m., it is a message as well.

Some years ago the management of a concern in England e-mailed the message that anyone who did not receive a notice before (date/time) would be fired. In 2006, RadioShack Corporation used email to inform 403 employees that their jobs had been eliminated. [262] But what when the email arrives at a different online identity address? Or is delayed by server breakdown? Or is deleted by the spamfighter?
The large amount of e-mails that is waiting for you after your vacation each have a different value than an e-mail that <ploink> comes in when you are online and working. A reaction to an invitation is useful as long as you are not already underway to your destination. A reaction to an advertisement is of value as long as you do not have the offered/requested item. As soon as one is provided from time t, the content of the message as well as the medium is not useful and therefore useless.

I have noticed that the young (in order to come to a choice/decision/result) set deadlines for themselves for something/someone to '*NOW!*' and often they send a question into cyberspace. They shoot pellets at everything that moves and each reaction is sniffed at and checked out for value/usefulness up until a certain moment; then it is no longer useful or even becomes undesired. From that moment on they start to ignore them. In certain situations they even physically close off reception, e.g. by logging out when they are online (to subsequently log-in again under a different nickname [= identity]) or by using some e-mail address or mobile phone number that they no longer use after the deadline (after some months these automatically die a virtual death by automatic reset).

Time is the Message

Earlier, I quoted Wiener who in mechanics views the factor time as a component of feedback, the timer of the current electronics, which in a model-based control of a message anticipates whether a corrective action is desired or necessary. Also, I referred to Bateson who describes a momentum, namely an event (in time) in the work behaviour of communication technicians, where one event instinctively represents another. In itself this also indicates that time is a message. Because of this development of telecommunication contacts, the ensuing increased speed of message traffic and the ability to manipulate it (at the moment of arrival, mailing and/or perusal) time as a message has begun to play a role.
For that matter, time also has a relation with identity. The body, the space you enter and the time (period) in which you live, play fundamental roles in the process of forming your identity. In puberty you learn to discover yourself in interaction with others, at home, at school and elsewhere.

---

[262] RadioShack lays off 403 via e-mail. Dallas News Online, 31-08-2006. WWW:
<http://www.dallasnews.com/sharedcontent/dws/bus/stories/083006dnbusradioshack.30c5bc1.html>.

The result of this perception is equality between the person on the one hand and his personality (identity) on the other hand. You express that personal identity, which is inextricably bound with your body. In the past the body was limited in its comings and goings, but changes in mobility, communication, lifestyle and leisure activities broaden the possibilities to give shape to yourself. At this moment, all our senses can be extended by technology, so in a fundamental way human can remote sense. It is no longer limited to your own physical environment. You can now test your individuality on the Internet, day and night, worldwide, multichannel via all media and in every culture, thus constantly developing it. If at a certain time you want to be somebody and want to have the related identity, then you just make yourself! And if the new identity does not suit you, you adjust it online. The Wannabe wannado. Unlike mobile telephony, where identity is linked to the device (via the so-called SIM-ID card), the Internet technology makes it possible to assume any number of different identities any time, which are used in order to enter into multiple relationships.

Emerging Technologies

The phrase 'Web 2.0' refers to a supposed second generation of Internet-based services – such as social networking sites, wikis, communication tools, and so called 'folksonomies' [263]– that emphasise online collaboration and sharing among users. Web 2.0 is rich in local knowledge, but also capable of linking to things in the outside world.[264] However, Web 2.0 sounds as a buzzword to hopeful entrepreneurs: it has the sweet smell of money about it. Web founder Sir Tim Berners-Lee is not big on the term and calls for calm. [265] He has some serious doubts whether Web 2.0 is at all different from the original Web (1.0). The 'web of human ideas' is served by the hypertext Web but the Semantic Web is about data and helps with machine analysis and automatic identification of the user (customer). It is said that people collaborate and share information online in a 'new' way and this clearly illustrates what 'Sir Tim' is talking about. It seems to be old ideas parading as new ones. Berners-Lee points out that in the past scientists have been trained to do things top down. In the business world projects are often the boss' vision made flesh. Not only the Internet but also the business models underlying the Internet have changed radically since the Dot.com crash. (Cassidy) [266] Socialising the Internet to commodity by making it a "natural" extension of the real world is really a step forwards.

Summary Chapter 4 'Social Aspects of Virtual Identity'

In this chapter the relation between media, time, message, society and identity is explored. The telecommunications technology has lowered barriers in our association with each other. Manners are simple, everyone is 'you', and there is no etiquette and little hierarchy. It is unisex all the way, without class and race difference. The social impact of identity management in relation to time and media is enormous. The Internet changed the traditional models of authority (social, intellectual, hierarchical) and this has lead to a new social order.
Time is the key to the feedback element of communication, so in this speedy century 'time is the message'. In the next chapter, the forms and varieties of virtual identities will be discussed.

---

[263] A folksonomy is an Internet-based information retrieval methodology consisting of collaboratively generated, open-ended labels that categorise content such as Web pages, online photographs, and Web links. A folksonomy is most notably contrasted from a taxonomy in that the authors of the labelling system are often the main users (and sometimes originators) of the content to which the labels are applied. The labels are commonly known as tags and the labelling process is called tagging. WWW: <http://en.wikipedia.org/wiki/folksonomy>.

[264] T O'Reilly, 'What is the Web 2.0? Design Patterns and Business Models for the Next Generation of Software' (2004); WWW: <http://www.oreilly.de/artikel/web20.html>. The term 'Web 2.0' was coined by a business conference organiser, to describe the subject matter of their post-dot com crash. Tim O'Reilly and Dale Dougherty formulated their idea of a new web concept by examples of change and laid out the seven basic principles of what distinguished Web 2.0 from Web 1:

- The Web As Platform
- Harnessing Collective Intelligence
- Data is the Next Intel Inside
- End of the Software Release Cycle
- Lightweight Programming Models
- Software Above the Level of a Single Device
- Rich User Experiences.

[265] Director of the World Wide Web Consortium; WWW: <http://www.w3.org/People/Berners-Lee/>. 'Berners-Lee calls for Web 2.0 calm'. *The Register*, 30-08-2006. WWW: <http://www.theregister.co.uk/2006/08/30/web_20_berners_lee/>.

[266] J Cassidy, *Dot.com*. (Harper Collins, 2002).

5 Forms, Varieties and Valuables of Virtual Identities

'It is the chiefest point of happiness that a man is willing to be what he is.'

Erasmus (1466-1536) Dutch humanist

Introduction

Is a virtual identity virtual? Is it any less real than the identity we put on when we apply for a job? Is it less real than the maiden body that has become the primary canvas on which girls express their identities, insecurities, ambitions, and struggles?
This suggests that identity is not fixed, that identity only exists in response to external stimuli. Given the definition that a Virtual Identity is the representation of an identity in a virtual environment, consisting of a property of objects that allows those objects to be distinguished from each other, it can and will exist independently from human control and may also (inter)act autonomously in an electronic system. But each virtual identity is surrounded by its own electronic environment, online or offline, in a local computer or in a global connected network. This chapter gives an overview of today's most common practices of virtual identities. It highlights both the positive and negative effects of this phenomenon.

Environment of Virtual Identities

A virtual identity can exist both in an offline and online electronic environment, fully separated or initially connected. It can exist independently from human control and can (inter)act autonomously in an electronic system. That system can be offline (such as a standalone game console) or online (connected to a network such as the Internet).

Intelligent Agents

Most virtual identities in electronic environments are facilitated by 'intelligent' software agents. A software agent is a software package that carries out tasks for others, autonomously without being controlled by its master once the tasks have been delegated. The "others" may be human users, business processes, workflows or applications. A basic software agent stands on three pillars, three essential properties: autonomy, reactivity, and communication ability. Each agent has also its own roles and responsibilities. The notion of autonomy means that an agent exercises exclusive control over its own actions and state. Reactivity means sensing or perceiving change in their environment and responding. Even the most basic software agents have the ability to communicate with other entities: human users, other software agents, or objects. (Figure 25)

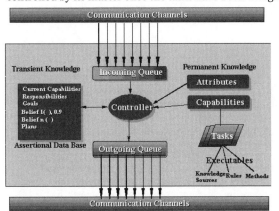

Figure 25: Anatomy of an Intelligent Agent [267]

An agent has both transient (the active workspace) and permanent knowledge (permanent means intrinsic in this context; it stays stable from one execution cycle to the next). It has a controller (representing its head) which can be given alternative behavioural characteristics (e.g. stimulus-response-like, actor-like, or blackboard-like). An intelligent agent performs tasks which have both declarative and procedural components. It can use alternative reasoning strategies, including belief

---

[267] F Farhoodi and P Fingar, 'Competing for the Future with Intelligent Agents' (*Distributed Object Computing Magazine*, 1997) WWW: <http://home1.gte.net/pfingar/agents_doc_rev4.htm>.

management. An agent's repertoire of tasks represents its capabilities. Each task can have its procedural "how to do" component represented as rules, methods, or knowledge sources (rule sets). In computer science, a multi-agent system (MAS) is a system composed of several agents, capable of mutual interaction. The interaction can be in the form of message passing or producing changes in their common environment. The agents can be autonomous entities, such as software agents or robots. When artificial intelligence technology is used, the software agents are called 'intelligent agents'. MAS can include human agents as well. Human organisations and society in general can be considered an example of a multi-agent system.

The behaviour of the electronic agents is restricted to the rules in the software code. Within this framework the tasks that the agents can perform and how the tasks are performed can be modified. Dependent on the level of added artificial intelligence in the software this modification can be dynamic and user centric, due to learning or the influence of other agents. In advanced games like 'World of Warcraft' (WoW) the non-playable characters (NPCs) are 'coded' to act and react to the actions of the opponent (human) player. After several rounds of struggle the agent would modify its behaviour in the next round by calling on other agents to help it out.

Figure 26: World of Warcraft: Most players at the battle field are artificial players, called 'agents'. They fight both independently and group-wise, depending on the design of the game software. [268]

Use and Users of Virtual Identity
This subchapter gives an overview of the various offline and online shapes of virtual identities.

Use of Virtual Identity

Online Conversation Modelling
A significant new and emerging field of applying virtual identities through intelligent agent technology is the modelling of conversation specifications and policies as positive/negative permissions and obligations, e.g. in mediation and arbitration cases, as a substitution of regular court procedures. Until now the agents mostly assist in document construction and software integration.
At the user-interface, agents work in conjunction with compound document and web browser frameworks and document management tools to select the right data, assemble the needed components, and present the information in the most appropriate way for a specific user and situation. Behind the scenes, agents take advantage of distributed object management, database, workflow messaging, transaction, web, and networking capabilities in order to discover, link, manage, and securely access the appropriate data and services.

Kagel and Finin [269] argue that Broersen et al use agent types to store solve conflicts between beliefs, obligations, intentions and desires.[270] The agent types are determined by their characteristics, namely

---

[268] Source: WWW: <http://www.lirmm.fr/~beurier/research.html>.

[269] L Kagal and T Finin, 'Modeling Conversation Policies using Permissions and Obligations'; WWW: <http://ebiquity.umbc.edu/_file_directory_/papers/92.pdf>.

[270] J Broersen, M Dastani, J Hulstijn, Z Huang, and L van der Torre. 'The boid architecture conflicts between beliefs, obligations, intentions and desire' (2001); WWW: <http://www.cs.uu.nl/~broersen/Papers/agents2001.ps>.

social (obligations overrule desires), selfish (desires overrule obligations), realistic (beliefs overrule everything else) and simple-minded (intentions overrule obligations and desires). Kagel and Finin state that within their framework conflict occurs basically between permissions and prohibitions, between obligations and prohibitions, as well as between obligations and dispensations. In order to resolve these conflicts, their framework includes meta-policies specifically setting the modality precedence (negative over positive or vice versa) or stating the priority between rules within a policy or between policies. Broersen et al approach the conflict resolution from the agent's point of view, whereas Kagel and Finin try to resolve conflicts in policies within the environment (and not within agents themselves). They believe that Broersen's approach or something similar could be used by agents after conflict resolution is provided by our framework as the enforced policies may conflict with the agent's internal beliefs, desires, intentions, obligations, prohibitions, and permissions.

Reasoning about beliefs, obligations, intentions and desires has been discussed in practical reasoning in philosophy [271], and its formalisation to build intelligent autonomous agents has more recently been discussed in qualitative decision making in artificial intelligence.
Broersen et al argue that on closer inspection each of these four concepts (i.e. beliefs, obligations, intentions and desires) consists of related (though often quite distinct) concepts, for example, knowledge and defaults, prohibitions and permissions, commitments and plans, wishes and wants. All these concepts are grouped into these four classes due to their role in the decision making process: beliefs are informational attitudes – how the world is expected to be – obligations and desires are the external and internal motivational attitudes, and intentions are the results of decision making. This results in their BOID architecture.
The question has been raised why norms are usually not implemented explicitly in computer systems. An easy answer is that computer programmes already model 'ideal' behaviour. They must never violate the rules, just as they must never fail. This objection can be countered by Dignum's argument that obligations can be violated, because agents are autonomous. [272] In a typical example, an agent has a desire to do otherwise and the desire is stronger than the obligation. Even social agents can violate their obligations if they intended earlier to do otherwise and are not open-minded enough to reconsider this intention. Finally, in order to deal with conflicts among norms, an agent must be able to drop some obligations in favour of others.
Types of middle-agent are mediator agent and matchmaker agent. (Figure 27) They are used in e.g. electronic markets and in multi-agent architectures for distributed learning environments.

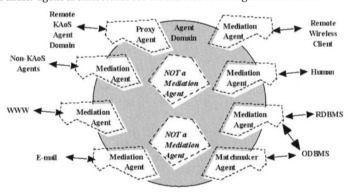

Figure 27: Mediation Agents, Proxy Agents, and the Matchmaker (NASA) [273]

Online Chat

Internet Relais Chat (IRC) is one of the earliest (1988) online multi-user chat systems, where people meet on Channels (rooms, virtual places, usually with a certain topic of conversation) to 'talk' in groups, or privately. There is no restriction to the number of people that can participate in a given discussion, nor to the number of channels that can be formed on IRC. Identity on IRC is established primarily by the use of an alias. An alias on IRC is much more diverse than Quake, as on IRC there is no object or point other than purely chatting to people. The object on IRC is to converse, to get to know people of similar interests and chat as long as you want. IRC is infinitely more complex than gaming in terms of the formation of identity. It is essentially structure less and allows a much broader range of identity types and alternatives.

---

[271] ME Bratman, *Intention, plans, and practical reason.* (Harvard University Press, Cambridge 1987).
G. von Wright. *Practical Reason.* (Oxford, 1983).
[272] F. Dignum. 'Autonomous agents and norms'. *Artificial Intelligence and Law*, 7:69{79, 1999.
[273] NASA, Mediation Agents; WWW: <http://ic-www.arc.nasa.gov/projects/aen/IA.html>.

Today's chat systems are extended with more features, such as graphics, actions and Peer to Peer (P2P) file transfer. They are also more equipped to personalise the user's virtual identity.

> Eccky is a virtual baby whom two chatters bring together on the (cyber)world. On the basis of their own individuality, behaviour and appearance arises a unique Eccky which appears under its own name in the buddy list. This virtual kiddy grows up in six days to an adult person. The chat-intelligence grows with to an unprecedented level (as yet). [274]

Online Intelligent Agent

It can be a virtual drummer of an orchestra in the virtual environment or a personal assistant in navigation through the virtual worlds.[275] [276] It also can be a virtual merchandiser who wants to sell you a flight ticket. More and more intelligent agents are used in the trading sector, mostly in the continuous transaction environment. (Figure 28)

Figure 28: Software Agents (Source: Fingar)

Computer networks like the Internet provide a novel way to arrange social contacts and to perform commercial transactions like selling consumer goods. A new tool to facilitate these (trans)actions is the artificial intelligent agent.

An intelligent agent usually operates as a part of a community of cooperative problem solvers, including human users. Each agent has its own roles and responsibilities. Fingar [277] points out that these multi-agent systems are essential in order to open digital marketplaces – especially considering the multiple dynamic and simultaneous roles a single company may need to play in given market sessions – because a manufacturer may need to be a broker, provider and merchant, simultaneously.

Table 1: Examples of Use of Intelligent Agent Technology

Workflow management.	Data filtering and analysis.	Negotiation.
Personal assistance / task delegation.	Bandwidth management.	Tutoring.
Condition monitoring and notification.	Resource scheduling.	Communications.
Collaborative systems integration.	Customization.	Arbitration.
Collaborative filtering.	Simulation and gaming.	Data mining.
Supply chain optimization.	Production control.	Profiling.

Behaviour of intelligent agents (the tasks that they perform and when/how the tasks are performed) can be modified dynamically, due to learning or the influence of other agents. A selling agent that just lost several rounds of sales would certainly modify its behaviour in the next round by calling on other agents to help it out. It is envisaged, that an intelligent agent will e.g. be able to find the best possible deal for a customer and then to perform the deal autonomously in the name of the customer. In most situations the customer is unknown with the dealing menu of the agents but having signed the end user agreement in advance the customer is obliged to the act. 'People are adaptive problem solvers and can deal with new situations "on their feet" and handle incomplete, inconsistent information in real-time, usually making good business decisions based on judgment gained from experience. This is precisely what computers must be empowered to do in eCommerce', argues Fingar. (Figure 29) He states that using object-oriented intelligent agents in eCommerce is neither science fiction nor should it be off putting. Thus, agents – built by Artificial Intelligence software – are not novel, but computer programmes that perform legal acts autonomously are uncommon, and

---

[274] Eccky; WWW: <http://www.eccky.com/>.

[275] M Kragtwijk, A Nijholt and J Zwiers, (2001): 'An Animated Virtual Drummer'; WWW: <http://www3.cs.utwente.nl/~anijholt/artikelen/icav_kr2001.pdf>.

[276] B van Dijk, R op den Akker, A Nijholt, and J Zwiers 'Navigation Assistance in Virtual Worlds' (2001); WWW: <http://www.informingscience.org/proceedings/IS2001Proceedings/pdf/ VanDijkEBKNavig.pdf>.

[277] P Fingar, 'Intelligent Agents: The Key to Open eCommerce; Building the next-generation enterprise'. WWW: <http://home1.gte.net/pfingar/csAPR99.html>.

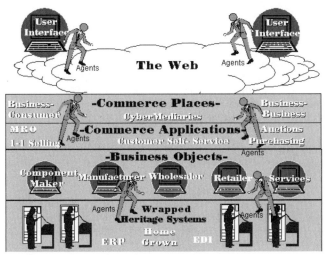

they are as yet not legally supported. For example, the legal status of contracts closed by intelligent agents is still unclear and there is a need to adapt laws to commercial transactions by way of intelligent agents. [278]

Figure 29: eCommerce Using Object-Oriented Intelligent Agent Architecture (Source: Fingar)

Recently the (Dutch) Computer Law Institute in Amsterdam started a study of intelligent agents and the law, as questions arose such as: 'Should intelligent agents be granted legal personality?' and 'What is the legal validity of contracts that have been closed by intelligent agents?' [279] It can be argued that these agents are not significantly divergent from legal persons like a foundation, a partnership or a company. This will be discussed in chapters 8 and 9.

Online Search Agent

Intelligent search agents may be equipped with features that enable them to effectively search the Internet. Normally these agents are programmed with respect to the status of exclusion clauses (like respecting the intellectual property rights) often found on Websites. To find information on a particular subject, a search agent may need to search exactly those web pages that are labelled as 'unsuitable for bots' by a no-robots clause or have some form of General Terms and Conditions. Whether these no-robots clauses apply to search agents as well and what the presence of terms and conditions on a Website entails for search agents is as yet unclear. [280]

Online Instant Messaging (IM) Agent

In our 24x7 global economies one cannot always be available for communication. If someone wants to ask something from another user or likes to provide some help or information to the other user when the other user is not available (offline), the communication by instant messaging can be automated by the concept of custom chat bots. The idea is to create a chat bot [281] to collect or provide information to the other side through chatting whenever the other side is available. Examples exist of 'agents' as virtual identities in MSN, such as requesting your bank balance.[282]

---

[278] E.g. Intelligent Software Agent Bibliography 1995; WWW: <http://ils.unc.edu/gants/agentbib.html>.

[279] Intelligent Agents and the Law; WWW: <http://cli.vu/onderzoek/agents/index.php>.

[280] ML Boonk, DRA de Groot, FMT Brazier and A Oskamp, 'Agent Exclusion on Websites', in *The Law of Electronic Agents* (2005) pp 13-20. WWW: <http://pubs.cli.vu/en/pub238.php>.

[281] Bot is an abbreviation from Robot; WWW: <http://www.computer-dictionary-online.org/?q=bot>. A "bot" refers to:
- Internet bot, a type of computer program to do automated tasks;
- Computer game bot, a computer controlled player or opponent (see the NPC in the online game World of Warcraft');
- Botnet, a network of 'zombie' computers used to do automated tasks such as spamming or reversing spamming;
- Botsourcing, the assignment of tasks to an automated, or intelligent, programmed agent;
- PhotoBot, a photo image of a face moving;
- Chatterbot, a computer program designed to simulate an intelligent conversation with a human.

[282] S Olsen 'Instant chat bot comes back to life. *CNET News.com*, Published: 20 01 2003. WWW: <http://news.com.com/Instant+chat+bot+comes+back+to+life/2100-1023_3-982670.html>.
Dutch Postbank started 18-09-2004 with automatic dialogue account balance information via MSN Messenger. Other IM agents are: Encarta(encarta@botmetro.net) accesses the extended multimedia encyclopedia of Encarta Winkler Prins; ESP Billy (espbilly@msn.com) is related to MSN Space to watch together a movie with one of your contacts; Smarterchild AOL (smarterchild@hotmail.com); Spleak (spleak@hotmail.com); TongYang securities (msn013@myasset.com); Crystal Ball (crystalballbuddy@hotmail.com); Inside Messenger (chat@insidemessenger.com); IM Local –Yellow Pages US (imlocal@msn.com); Chatman' (blabla@chatman.nl); Yellow Pages Canada (yellowpagesqa@hotmail.com).

The creation of bots is a very simple and straight forward software building procedure. Two types of bots are popular, i.e. Q&A bot & Helper bot. As it is clear from QA name that this bot will have some questions & answers to be delivered, based on the responses of end user. To create a Q&A bot, you add a question in the bot & then add keywords based on which next question will be asked. When a user responds to a question, its response is matched with the keywords & then the next question is asked. The creation of a Helper bot is same as a Q&A bot however the Helper bot has a different flow then the QA bot. When responding to a user the Helper bot checks for the best possible response. To deploy the bot in a professional environment with messengers like MSN, Yahoo, Google Talk, ICQ, AOL some additional converging software is needed.

WebBot – Virtual Assistant on the Web

WebBots (also called automatic avatars) are virtual assistants that guide the user through the Internet. A WebBot is usually individually tailored to suit your image. WebBots are similar to software assistants and, like these, are versatile tools with 'personality'. Integrated search engines change into characters that interact with users. E.g. "Anna" of Ikea's Website shows emotions (like smiling). Depending on the programming used, these more or less intelligent WebBots keep the user company, providing advice and helping him to conduct searches and sort through information. Their special look helps to give a Website a personal note.

Botnet(work)

A computer that has bot software installed – for example through a malicious Website – is called a 'zombie'. A zombie PC can be used to carry out an (e.g. criminal) activity without the owner's knowledge. A network of zombies is referred to as a 'botnet'. The zombies can be controlled remotely by the attacker, who can send commands while the PC owners are oblivious to what's happening.

Botnet is a jargon term for a collection of software robots, or bots, which run autonomously. This can also refer to the network of computers using distributed computing software. While the term 'botnet' can be used to refer to any group of bots, such as IRC bots, the word is generally used to refer to a collection of compromised machines running programmes, usually referred to as worms, Trojan horses, or backdoors, under a common command and control infrastructure. A botnet's originator (aka 'bot herder') can control the group remotely, usually through a means such as IRC, and usually for nefarious purposes. Individual programmes manifest as IRC 'bots'. Newer bots can automatically scan their environment and propagate themselves using vulnerabilities and weak passwords. Generally, the more vulnerabilities a bot can scan and propagate through, the more valuable it becomes to a botnet controller community.

> Botnets have become a significant part of the Internet, albeit increasingly hidden. Due to conventional IRC networks taking measures and blocking access to previously-hosted botnets, controllers must now find their own servers. Often, a botnet will include a variety of connections, ranging from dial-up, ADSL and cable, and a variety of network types, including educational, corporate, government and even military networks. Sometimes, a controller will hide an IRC server installation on an educational or corporate site, where high-speed connections can support a large number of other bots. Exploitation of this method of using a bot to host other bots has proliferated only recently, as most script kiddies (young computer nerds) do not have the knowledge to take advantage of it.

> Botnets are often rented out by their owners, called bot herders, to relay spam and launch phishing scams to steal sensitive personal data for fraud. Botnets have also been used in blackmail schemes, where the criminals threatened online businesses with a denial-of-service attack on their Website to extort money. Several high-profile companies were hit by variants of a 'worm' that installed itself on company computers connected to the internet so that they could be accessed by hackers to send spam or to disrupt the companies' networks. [283]

Virtual Assistants of the Criminals on the Web

Criminals use the Internet to exploit their activities in this new efficient environment. The diagram in figure 30 illustrates how a botnet is created and used to send email spam. A botnet operator sends out viruses or worms, infecting ordinary users' computers, whose payload is a trojan application – the bot. The bot on the infected PC logs into a particular IRC server (or in some cases a web server). That server is known as the command-and-control server (C&C). A spammer purchases access to the botnet from the operator.

---

[283] Botnets. See: WWW: <http//en.wikipedia.org/wiki/Botnet>.

The spammer sends instructions via the IRC server to the infected PCs, causing them to send out spam messages to mail servers.

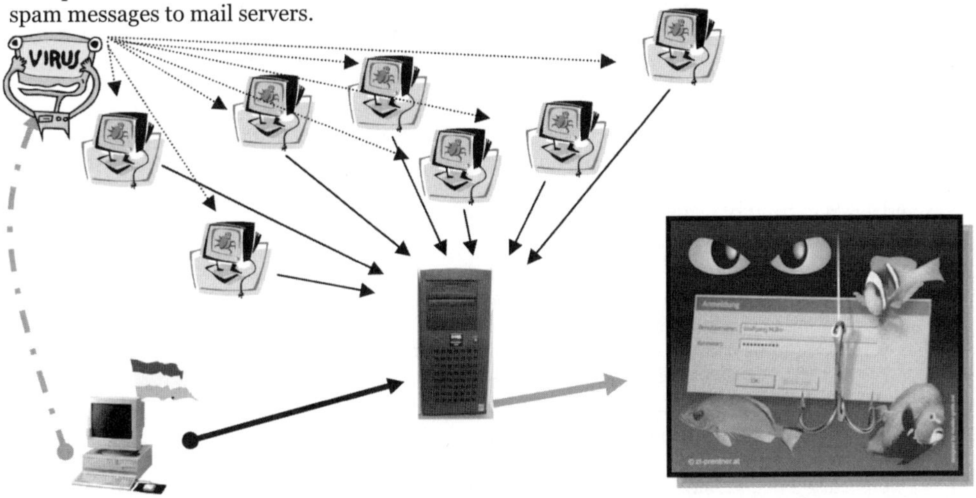

Figure 30: A diagram of the process of using zombie (virus-infected) computers to send spam (source: Dutch police)

Bots are a serious computer security problem, and law enforcement seems to just be catching up to it. In 2005 a turf war erupted on the Internet between competing hackers trying to hijack computers running Microsoft's Windows 2000 operating system in order to turn them into 'zombie PCs'. Several botnets were found and removed from the Internet. In October 2005 the U.S. authorities announced the first bot-related arrest, just prior to police in the Netherlands said that three men suspected of hijacking about 1.5 million node PCs had been arrested;[284] and the Norwegian ISP Telenor disbanded a 10,000 node botnet.[285] Large coordinated international efforts to shut down botnets have also been initiated. [286] It has been estimated that up to one quarter of all personal computers connected to the internet are part of a botnet.[287]

Offline Games
Offline games, like SIMS, can be played without accessing the Internet. These games are comprised of PC-packaged games for playing on personal computers, video game consoles or console games as well as handheld video games, but when the console or PC is connected to the Internet the games often can have some interactive contact for help functions, community access or level upgrading. The software is typically made available to consumers on a disk or a cartridge. Although some offline games have network features that allow users to play on a peer-to-peer basis on a local area network or through a publisher's server, user and individual game data are not stored on the servers. Offline games – or games that you play alone – often offer less elements than online games, but also have virtual identities.

Online Games
Online games have at least one component that must be played by accessing the Internet. Online games are predominantly played on personal computers, although they can also be played on various other devices. The main part of the game software operates on network servers to which end users have no access, and user and individual game data are stored on the servers. Massive multiplayer online (role-playing) games (MMORPG's) require different research methods than other games. Virtual identities were used as the player's creature, such as in the game play 'Black & White', in which the game's complex artificial intelligence (AI) is enabled to train a pet of sorts to do almost anything. Artificial creatures (persons) are also used in popular online 3D first person gaming, but the different roles that people take on in online role-playing games (ORPG) are still structured.

---

[284] G Keizer, 'Dutch Botnet Suspects Ran 1.5 Million Machines'; (2005) WWW: <http://www.techweb.com/wire/security/172303160>.

[285] J Leyden, 'Telenor takes down 'massive' botnet. Clients are still zombies' in: The Register (2004); WWW: <http://www.theregister.co.uk/2004/09/09/telenor_botnet_dismantled/>.

[286] J Leyden, 'ISPs urged to throttle spam zombies. International clean-up campaign' in: The Register, (2005); WWW: <http://www.theregister.co.uk/2005/05/24/operation_spam_zombie/>.

[287] BBC News, 'Criminals 'may overwhelm the web'' (25-01-2007); WWW: <http://news.bbc.co.uk/1/hi/business/6298641.stm>.

Unless the programme is altered, the players are forced to take on a super-masculine character. Ten years ago, in 'Quake', the appearance of this character was set by default except for the colour of his combat gear. Today, in 'World of Warcraft' the characters can be fully personalised. Old or new, as both games are a graphic, action-orientated medium, it follows that this is also how identity is performed. 'World of Warcraft' is a war game, and as such, players establish virtual identities to suit the medium. 'Quake' was primarily played by young male secondary and tertiary students, as is 'World of Warcraft'. The game is aimed towards this demographic and as such caters to their fantasies. Role-playing games offer a chance to take up the super-ego representation of the big action movie star. The virtual identity an online role-playing game offers is simply an extension of the identity that has been promoted again and again in front of us in the mass media. The computer game is an arena for entertainment.

World of Warcraft

The Massive Multiplayer Online Role Playing Game 'World of Warcraft' (WoW) is an example of the way in which a virtual identity ('NPC' [288]) is forced upon unsuspecting or apathetic players. The players are given an opportunity to become their super-ego. They become the action hero from the cinema. This is exemplified by the lack of real means to choose the virtual identity.

The main character is a white male. The point must also be made that this is the target audience. The software was not written as a communication device, but rather as a multimedia experience. The game caters for a specific type of virtual identity. The fragmented identity paradigm is applicable in this situation. An online role playing game presents certain stimuli, granted they are not fluid and infinitely adaptable such as IRC, but it still presents stimuli. The player reacts to these stimuli by forming a `virtual' identity that is appropriate. While the player is involved in the game, he is this identity. The fact that the world the player is in does not exist is superfluous. The player is interacting with others, had conceived a certain identity, and while he is playing, is that identity.

Online Presentation

Wood & Smith [289] states that Internet technologies offer the possibility to get grip on your identity. In the virtual world we can decide which part of our identity will be presented to others.

> 'Websites, that are visibly 'under construction', are not only the pages but the authors themselves.'
> Chandler [290]

Chandler states: The *content* of personal home pages can be recognised as drawing on a palette of conventional paradigmatic elements, most notably: personal statistics or biographical details; interests, likes and dislikes; ideas, values, beliefs and causes; and friends, acquaintances and personal 'icons'. Creating a personal home page can be seen as building a virtual identity insofar as it flags topics, stances and people regarded by the author as significant (as well as what may sometimes either be 'notable by its absence' or 'go without saying'). Turkle[291] notes that in a home page, 'One's identity emerges from whom one knows one's associations and connections'.

Online Communities

Every person who communicates in virtual and social environments owns a virtual identity, which sometimes may be identical to real identity. As mentioned before, our lives are a stage on which we are playing out roles that we have learnt from birth onwards. This virtual identity finds a way to express itself by existing on the Internet. Transferring feelings, thoughts and links to online communities is important. 'Show me what your links are, and I'll tell you what kind of person you are'. (Miller) [292] Where such links are to the pages of friends, or to those who share one's interests, can be seen as involving the construction of a kind of 'virtual community' by home page authors. (Rheingold) [293]

---

[288] 'NPC' in a computer role playing game or other computer game is the acronym for Non-Playable Character, such as traditional personages. It is a computer-manipulated personage, that performs as enemy or friend and is programmed (often by artificial intelligence) to follow specific patterns of (human such as) behaviour in relation the behaviour of the connected party. NPC's are increasingly used in Multiplayer Online Role Playing Game's. A NPC is an artificial person, in fact a juristic person. WWW: <http://en.wikipedia.org/wiki/Non-player_character>.

[289] AF Wood and MJ Smith, *Online Communication. Linking Technology, Identity & Culture* (Lawrence Erlbaum. London 2005).

[290] D Chandler, 'Personal Home Pages and the Construction of Identities on the Web' (1998); WWW: <http://www.aber.ac.uk/media/Documents/short/webident.html>.

[291] S Turkle, *Life on the Screen: Identity in the Age of the Internet*. (Weidenfeld Nicolson 1996) p 258.

[292] H Miller, *The Presentation of Self in Electronic Life: Goffman on the Internet* (paper for Embodied Knowledge and Virtual Space conference, London 1995) WWW: <http://www.ntu.ac.uk/soc/psych/miller/goffman.htm>.

[293] H Rheingold, *The Virtual Community: Finding Connection in a Computerised World*. (Minerva, London 1995).

Ultimately, online identity cannot be completely free from the social constraints that are imposed upon in the real world. As Westfall discussed, the idea of truly departing from social hierarchy and restriction does not occur on the Internet (as perhaps suggested by earlier research into the possibilities presented by the Internet) with identity construction still shaped by others. [294] Westfall raised the important, yet rarely discussed issue of the effects of literacy and communication skills of the online user. Indeed, these skills or lack thereof, shape one's perception in the online community in a similar way to that of the physical body in the 'real world'.

There is a different facet to online identities: those personae that are adapted, imposed, or suggested by computer games, email, chat rooms, blogs, instant messaging software, or other online sites. Concerns regarding virtual identity revolve around the areas of misrepresentation and the effects between the online and offline existence. The idea of every user's ability to self-portray has resulted in much discussion about the validity of online relations. Sexual behaviour online (and the ability of paedophiles to obscure their identity) provides some of the most controversial debate, with many being concerned about the predatory nature of some users. However, its possibility to highlight the interplay between social circumstances and how people choose to construct their sexual and gender identity in response to this can be analysed through the online opportunities that the Internet presents. Therefore the opportunities, fluidity, and problems of the brave new world of virtual identity creation should be examined.

Online Dating

People around the globe are harnessing the connecting power of the Internet, to find anything from friendship to marriage. The nature of the Internet means people can be connected to each other regardless of geographical boundaries. The Internet gives the power of anonymity to the individual, creating both safeguards and new dangers. Several online dating sites include chat as part of their service. Examples are fri.com and Yahoo! Personals. On these sites a subscriber logs in, makes himself an online profile with virtual identity, and sends messages to potential mates who are currently online. This chatting system is very similar to ordinary chat rooms. Although there is a degree of social stigma connected with online dating services, results released by Match.com and DatingDirect.com revealed over half a million people in the U.K. visited these Websites. Dating sites and their chat rooms can be accessed all day, everyday, whereas in the non-virtual plane social activities must be undertaken at certain times. In addition, such dating services provide a cheaper alternative for social contacts than many activities can in the real world. (Burmaster)[295]

Online Interactive Broadcasting

Movies, cartoons, television series, advertisements, multimedia and computer games are all built around narrative structures. However, the potential of interactive narratives has not yet been fully explored by interactive television in comparison to its many uses in computer games and virtual environments. The virtual identity is used for authoring interactive experiences. With this approach the user is experiencing virtual environments through the eyes of a virtual identity, resulting in different interactive experiences within the same environment. For example, the user, being the main star in the scene, can individually replay an original movie. The BBC focused in 2002 on the transition mechanism and the ways it can be used to enhance interactive experiences. [296] It also proposed ways of using the approach with today's technology of interactive broadcasting.

Online Interactive Storytelling

Interactive storytelling is an initial virtual identity approach in the Cultural Heritage area that relates real identities to virtual identities, in terms of socio-psychological, gender and embodiment issues. The participant experiences the interactive story through the eyes of a virtual identity. Each virtual identity is defined by the knowledge about itself, by its perception about the environment and its virtual embodiment. Key elements are:
- Characteristics that a virtual identity is born with;
- Characteristics concerning the virtual identity's background;
- Embodiment of the virtual identity;
- Behavioural characteristics of the virtual identity.

H Rheingold, 'Multi-User Dungeons and Alternate Identities' in: *The Virtual Community: Homesteading on the Electronic Frontier*. (Harper Collins, New York) Chapter 5; Cited from WWW: <http://www.well.com/users/hlr/vcbook/vcbook5.html>.

[294] J Westfall, 'What is cyberwoman? The Second Sex in cyberspace', *Ethics and Information Technology*, no.2 (2000).

[295] A Burmaster, '21st Century Dating: The Way It Is accessed' (2005); WWW: <http://www.nielsen-netratings.com>.

[296] BBC Research & Development, Virtual Identities in Interactive Broadcasting (2002): WWW: <http://whitepapers.silicon.com/0,39024759,60032791p-39000576q,00.htm>.

For some years the interactive storytelling is also used in the so called cross-media. Connecting the cross-media components (ordinary television, interactive television, radio, mobile telephone (including SMS, billboards, outlet-displays, and all kinds of print) to the Internet will stimulate the use of virtual identities. Greeff and Lalioti carried out some experiments, demonstrated in Cyberstage with a surround-screen projection-based stereoscopic display system[297].

Online Worlds

Online worlds, also called 'Virtual Worlds', 'Synthetic Worlds' or 'Massive Multiplay Online' (MMO), let you create a character, a home, a pet, and a new personality if you wish. You can take on jobs, be part of a community, and have fun hanging out with others in an online world. Online worlds have a focus on social interaction. 'Active Worlds' has over 1000 unique worlds to immerse yourself in. Shop, play games, and hang out with others. Rick's Café, Pollen World, and Castles World are samples of worlds you can be a part of. 'Second Life' has a complex social and economic structure. You can fully customize your home and appearance, and all surplus can be sold. The so-called virtual 'Linden dollars' can be exchanged into real money. "Moove" is a 3D online world. Decorate your home, invite others to come over, or explore the world by visiting your neighbours.
Online worlds are not only a play ground and training wheel for adolescents, but will soon become training grounds for artificial intelligences. [298]

Online Machinema

Machinema [299], a film created by means of automated production techniques and called Computational Drama[300], is a combination of drama, new media and applied information technology. 'Broken Saints' is the first independent Internet project and it goes some way towards proving that there is more to the series' appeal than its initial medium. [301] Rooster Teeth's 'Red vs. Blue' is an online Machinema production created by independent Texan film-makers. [302] Throughout the series, new characters are added and an incredibly deep plot is uncovered. All together it's something you can watch again and again and never be disappointed. A mix of MMORPG and Machinema can be found in 'Make Love, Not Warcraft' by Kenny from the TV-movie 'Southpark'.[303]

Online Virtual World Movie

In January 2007, a man named Molotov Alva, disappeared from his Californian home. Recently, a series of video dispatches by a Traveller of the same name have appeared within a popular online world called SecondLife. Filmmaker Douglas Gayeton came across these video dispatches and put them together into a documentary of seven episodes. (Figure 31) [304]

Figure 31: Molotov Alva in SecondLife

Online Relational Networks

Each of us has at least one online identity that is manifest on the web. Many people invested much in creating and decorating their online identity by means of profiles et cetera. Others keep a blog page as their virtual identity, with links to all their sub-identities (homepage, friends, rss, nicknames, other blogs). An obstacle arises where identity comes in. When the web crusader leaves because a newer, cooler hangout spot comes along, all of his/her messages and profile information will be lost.

---

[297] M Greeff and V Lalioti, 'Interactive Storytelling with Virtual Identities' (2000); WWW: <http://www.makebelieve.gr/vl/Publications/IPT_EGVE_final.pdf>; and

C Jackson and V Lalioti, 'Virtual Cultural Identities' CHI-SA 2000, South African Human-Computer Interaction Conference 2000, University of Pretoria.

[298] BBC, 'Online worlds to be AI incubators' 13-09-2007. WWW: <http://news.bbc.co.uk/1/hi/technology/6992613.stm>.

[299] Machinema concerns the rendering of computer-generated imagery (CGI) using real-time, interactive (game) 3D engines and complex 3D animation software. WWW: <http//en.wikipedia.org/wiki/Machinema>.

[300] Avatar Body Collision (2002); WWW: <http://www.avatarbodycollision.org/index2.html>.

[301] Broken Saints (2001): Machine created interactive movie; WWW: <http://www.imdb.com/title/tt0451002/>.

[302] Red vs Blue (2003): Machine created interactive movie; WWW: <http://www.imdb.com/title/tt0401747/>.

[303] Episode 147 of Comedy Central's animated series South Park (aired on 04-10-2006). This episode is a parody of the popularity of the massive multiplayer online role-playing game World of Warcraft.

[304] 'My second life; the video diaries of Molotov Alva', movie by D Gayeton; WWW: <http://www.molotovalva.com/play.html>.

In previous years web services (thing finders such as Thinglink) and social network sites (such as Classmates.com, Linkedin, MySpace) enabled information storage in a kind of personal identity domain, allowing people to keep their contact information in a centralised place so that their contacts would all know when they have changed it. A point of discussion is whether identity building is restricted to self-identity building, i.e., can only you specify information about yourself? This however seems limiting, as in the real world we construct our perceptions of other people all the time. And in the virtual world, for some people, keeping track of all the things about them (articles, blog entries, etc.) could be a full-time job.

Online Self-service

New technological possibilities make a self-service development concept possible, in which all the parties in the information chain get (authorised) access to relevant data that is made virtually available depending on their wishes and demands. A virtual entity that acts as a link station (broker) is created. This entity supports both technically and procedurally the exchange of data between the different parties in the information chain. When these parties are connected to the link station the exchange between parties converts entirely automatically in the information chain. In the realisation of the self-service concept several modalities are possible, which may vary depending on the degree in which the data is stored centrally or locally. In the first case the entity also has a role as virtual information warehouse, in the second case it will be stored locally especially for the routing of data. The Dutch RINIS Foundation is an example of such a virtual information warehouse. [305]

Online Feelings and Satisfaction

Nicholson Baker's novel 'Vox' [306] focuses on the detailed content of a phone call between Abby and Jim, both visiting a sexual dating service, and talking about ways to pleasure each other online. In the following years creative students built scripts and toys in order to satisfy them from a distance. The consequences of these 'services' and 'tools' for social behaviour on the Internet were discussed by Holmes [307]: 'One of the differences is the intensity of the interface. Some ways of contact allows interaction to be reduced, for example, to a telephone conversation of hearing-voicing without immediately provoking an identity cri-sis. One of the most common defences of cybersex is now the argument that techno mediation allows for safe sex, AIDS-free interaction.'

\{{{{{{{{{{{{{{{{{{{{{(everyone)}}}}}}}}}}}}}}}}}}}}}] and the ladies twice ;-)

This is an online hug. It comes from the text of an IRC (Inter-Relay Chat) session, captured by Zaleski [308]. He states that this is 'a ghostly thing,' pale in comparison to a 'real' hug. Despite his high hopes for the spiritual possibilities developing in cyberspace, he is deeply disturbed by the fact that it is a disembodied realm. He posed the rhetoric question 'How can you have rituals without bodies?' Meredith Underwood quoted this text too and replied 'How can you have hugs? Well, this is how.' She is not sure just what about this online hug is supposed to be ghostly. Actually Underwood used it herself (minus the second part about the ladies and substituting the appropriate name for 'everyone') while chatting with a woman who was struggling through a number of life transitions. She thought it marvellous, hardly concerned at that moment with fine distinctions between the 'virtual' and the 'real.' And, more to the point, she felt hugged.

By thus inventing icons and emoticons such as this one, followed by avatars and flashes, online DIYs (do-it-yourselfers) find ways to convey gesture, expression, tone of voice, and other forms of embodied communication supposedly missing in cyberspace (Underwood) [309]; (Shaw) [310]; (Turkle)

---

[305] The RINIS Foundation (Institute for the Routing of (Inter)National Information Streams) is a Dutch organisation in which all the major public implementing agencies in the Netherlands work together, also with other countries and international parties. It is a network organisation which supports its members in all aspects of data exchange for the implementation of public duties. In addition to arranging for data exchange by means of standardised messages, RINIS also serves as a platform for consultation in which members conclude agreements with each other on data exchange. Within RINIS, data exchange takes place through sectoral access points, which relay queries and replies to and from the participating organisations. This is done using independently operating computers which exchange messages of a predefined format only. These RINIS servers ensure that messages are exchanged safely, enabling efficient and effective re-use of once-recorded data, in accordance with privacy law. WWW: <http://www.rinis.nl/ENGELS/index.htm>.

[306] N Baker, *Vox* (Vintage, 1992; Reissue edn 1993).

[307] D Holmes, *Virtual Politics: Identity and Community in Cyberspace* (Sage, London 1998) p 109.

[308] J Zaleski, *The soul of cyberspace: how new technology is changing our spiritual lives.* (Harper, San Francisco 1997) p 248.

[309] M Underwood, 'Wired Women; Lost (or Found?) in Cyberspace'. *Journal of Religion and Society*; V1/1(1999) p 11. ISSN: 1522-5658. WWW: <http://moses.creighton.edu/JRS/1999/1999-5.html>.

[310] DF Shaw, Gay men and computer communication: a discourse of sex and identity in cyberspace. p 134 / pp 132-45 in SG Jones (Ed) 'Virtual culture: identity and communication in cybersociety'. (Sage, London 1997).

[311]. It seems – like the blind and deaf – people replace the lacking senses in transferring information by a mix of other sensorial qualities.

Yet even beyond such creativity, there lies a plethora of cultural evidence attesting to what Deborah Lupton [312] calls 'the embodied computer/user'. Advertising, for instance, often represent PCs as an extension of the human body and ascribe human feelings to them. An intimacy has developed between us and our computers that is physical as well as emotional. 'We can now carry them about with us in our briefcases, and sit them on our laps,' Lupton observes. 'They take pride of place in our studies at home and our children's bedrooms'. And it works both ways: 'Our interactions with PCs 'inscribe' our bodies,' Lupton noted, 'so that, for example, pens start to feel awkward as writing instruments'. Underwood argues that cyberbodies are clearly not flesh and blood; they do not literally touch, skin to skin. 'Embodiment was not the most controversial subject I discussed with women online, but it was by far the most difficult to articulate. Generally, they talked of minds and souls reaching out to each other and connecting on the Web. But further conversation nuanced this account, suggesting an intense and tangible relationship between the bodies of computer/users, mediated by the technology and occurring somewhere in the interstices of cyberspace.'

Online Intimacy

The next step in a human contact, such as an intimate 'hug', was presented in Wired [313] with the 'Teledildonics', an article about people enriching their identity with remotely controlled sex toys. Again in 2007, Wired also presented the dual fun tools for online pleasure on both sides. [314] 'If you wanted to be a porn star,' said Emi, a spokeswoman for the manufacturer, 'but you didn't want to get an agent and a manager and all that, you could use our products online and build a fan base and get famous.' When you stroke the device with your hands, or lips, or whatever, the software captures the placement and pressure of each touch and embeds the signals into the video signal of a DVD player. During playback, those signals are translated and sent to another USB device, that reproduces the (in)decent acts. In a distance contact with these kinds of interfaces, from touch and taste to smell and 3D vision, the receiver's perception will be realistic. In combination with virtual worlds, one's imagination to come will reach reality, but the (pseudo-)identity will stay in virtuality.

Online Body

Wiener defined in his 'Control and Communication in the Animal and the Machine' three central concepts which he maintains are crucial in any organism or system: communication, control and feedback. Wiener postulated that the guiding principle behind life and organisation is information, the information contained in messages. As a governing mechanism, messages rely on feedback; different inputs lead to different outputs, and the computer essentially is a simulating machine that processes inputs. Following Wiener, others discussed this cybernetic phenomenon; that we have to think about this process as an evolution of the cooperation between man and machine or the vanishing of the boundaries between them, an evolution that is both a product of design and a process with its own dynamics. As Hayles points out, the illusion that information is separate from materiality could lead to a dangerous split between information and meaning, as well as restricting the space of theoretical inquiry.' [315]

Over the past six years, many online artistic projects have addressed issues surrounding the cyberbody and the relation between our physical bodies and digital technologies. Eduardo Kac's 'Time Capsule' [316] crossed the frontier between the body and technology invasion by turning his body into a 'site' hosting artificial memory by web-scanning parts of himself in Chicago, and then registering himself at a remote web-based animal identification database in Brazil that originally was designed for the recovery of lost animals. It was the first time a human being was added to the database as Kac registered himself both as animal and owner. The event was shown live on national television and on the Web. On the basis of this kind of project, one could make convincing arguments that we already have turned into technologically enhanced and extended bodies. Shaken-up by the actual developments of genetic engineering we are just beginning to understand the effect

---

[311] B Turkle, *Life on the Screen: Identity in the Age of the Internet*, (Simon and Schuster 1997) p 47 ff.

[312] D Lupton, *The embodied computer/user*, pp. 97-112 in M Featherstone and R Burrows (eds): 'Cyberspace/cyberbodies/cyberpunk: cultures of technological embodiment' (Sage, London 1995).

[313] R Lynn, 'Ins and Outs of Teledildonics', in: *Wired* 24-09-2004; WWW: <http://www.wired.com/news/culture/0,65064-0.html>.

[314] R Lynn, 'Teledildonics Takes a Step', in: *Wired* 19-01-2007; WWW: <http://www.wired.com/news/columns/0,72524-0.html?tw=rss.index> and WWW: <http://www.girlsr.tv/products/index.html>.

[315] WWW: < http://www.asc-cybernetics.org/> <http://pespmc1.vub.ac.be/Asc/ASCGloss.html>.

[316] E Kac, 'Time Capsule' (1997). WWW: <http://www.ekac.org/timec.html>.

that technologies have on our physical bodies. The cyborg as cyberbody is an unexplored field. Given the makeability one might speculate that the boundaries of our bodies will continue to dissolve and that the philosophic question 'Who am I?' will become less relevant in the future and will be replaced by 'Who or what can I be?'

Balsamo treated this disembodied look on the Internet more specifically in the variety of concrete forms of its appearance, such as the alternative 3D space of the virtual reality (VR). He varies the surroundings of the computer users, using technical resources as VR-glasses and pressure suits, in such a way that the 3D environment continuously adapts itself to the operations of the user. [317] VR does not represent (like a picture) the real environment, since it calls up no pre-existing reality but it results in an entirely new created reality without having to submit to physical law.

Victoria Vesna has developed 'Bodies INCorporated' [318], a site in which participants can construct new life-forms such as virtual identities or avatars. This shows great promise towards our body ('meat' in terms of the cyberpunk authors) no longer playing a role in the virtual space. However, the reality turns out to be different, for example when one examines the concrete behaviour in the virtual spaces. The possibility of making abstractions of the body does not mean, according to Balsamo, that one enters this space without a body. Dominating gender-thinking confirms the self-chosen representation of bodies in the virtual spaces. Even the interactions in virtual meeting places, where interaction is not based on text and on pictures, do not prove to be completely different from those that take place in the real world. (Stone)[319]

Plant prophesises that 'Masculine identity has everything to lose from this new techniques. The sperm count falls as the replicants stir and the meat learns how to learn for itself. Cybernetics is feminisation.

> *'The machines and the women mimic their humanity, but they never simply become it. They may aspire to be the same as man, but in every effort they become more complex than he has ever been. Cybernetic feminism does not, such as many of its predecessors, (...) seek out for women a subjectivity, an identity or even a sexuality of her own: there is no subject position and no identity on the other side of the screens. And female sexuality is always in excess of anything that could be called 'her own'. Woman cannot exist 'such as man', neither can the machine. Her missing pieces, what has never been allowed to appear, was her connection to the virtual, ... computers: they are the simulators, the screens, the clothing of the matrix, already blatantly linked to the virtual machine of which nature and culture are subprograms.'*
>
> Sadie Plant [320]

Online Death

The Internet and various mechanisms technological innovations that can be used by different people for different purposes. Cyberspace may attract psychotic persons by its virtual reality. It could be argued, though, that these people are psychotic in the real world and find a means to actualise their views in the virtual one.

> *'The case of a Maryland woman who had a business on the World Wide Web and that in electronic 'chat rooms' dealing with necrophilia asked several times to be tortured to death. (...) several men corresponded with her but stopped when they concluded she was serious'. Other people on the Internet tried to convince her to forget her intentions. The woman finally met a man who 'engaged in raw, sexual and violent conversations' with her by email. She travelled to meet this man, a computer programmer in North Carolina. After her husband notified the police that she was missing, her body was found with a rope that 'may have been the means of her death', buried by the home of the computer analyst. The man is being charged for homicide. The police found a letter left behind near her computer that said: 'if my body is never retrieved, don't worry. Know that I am at peace'.'* [321]

---

[317] A Balsamo, *Forms of Technological Embodiment: Reading the Body in Contemporary Culture*, in: M Featherstone and R Burrows (eds), 'Cyberspace/Cyberbodies/Cyberpunk : Cultures of Technological Embodiment'. (Sage, London 1995)

[318] Bodies INCorporated; WWW: <http://www.bodiesinc.ucla.edu>.

A Dempsey, 'Through the Window: New Media, Identity, and the Public Sphere'. (2005) WWW: <http://www.ibiblio.org/nmediac/summer2005/window.html>.

[319] AR Stone, *Sex and death among the disembodied: VR, cyberspace and the nature of academic discourse*. In: SL Star, (ed) 'The Cultures of Computing' (Blackwell, Cambridge 1995) pp 243-255.

[320] S Plant, *The Future Looms: Weaving Women and Cybernetics*, in 'Cyberspace / Cyberbodies / Cyberpunk' (1995) by Mike Featherstone and Roger Burrows (eds.), p 63 and in: 'Clicking In' by L Hershman Leeson, ed (Bay Press, 1996) p 132.

[321] D Struck and S Fern, 'A Cyberspace Fantasy Turned Fatally Real. Testing Limits of Cyberspace Led Maryland Woman to her Death', *The Washington Post*, November 3, 1996, pp. A1 and A22.

In LambdaMOO, Dibbel reports, a series of violent 'rapes' by one character caused a crisis among the participants, one that led to special conferences devoted to the issue of punishing the offender and thereby better defining the nature of the community space of the conference. Child porn and paedophiles are also reported in virtual worlds. This experience also cautions against depictions of cyberspace as Utopia: the wounds of modernity are borne with us when we enter this new arena and in some cases are even exacerbated. (Dibbel) [322]

Online Resurrection

'You only die once' but your online life continues forever (if you pay $9.99 for a three-year subscription). "Mylastemail" is an online service which allows you to leave – after your death – personal love messages, private expressions of appreciation, and words of encouragement for those you care about. You can log on at any time and edit your emails preparing your farewell. This keeps your messages relevant, up-to-date and even more personal. Stringent security checks ensure that your emails will not be delivered, or seen, until after your death. After departing, your final message from the grave can be of great comfort to the ones you leave behind. Now you need never leave anything unsaid. Your virtual identity is representing you.

Figure 32. Online contact with the hereafter? [323]

Online Transactions and e-Commerce

Most e-marketplaces, webstores and service providers use virtual identities to arrange the user characteristics and profiles. For example, eBay links the user account to Skype and PayPal. Like Amazon, many of the most colourful images are graphics supplied by commercial sites, with links to them. There are home pages that are essentially advertising bill-boards for products and services ranging from browsers to shoes to entertainment conglomerates. The construction of virtual identity through commercial symbols seems to be the presentation of the self as an advertising medium.

Online Property

With the rise of user-created property in virtual worlds, the development of copyright laws worldwide and the increasing success of alternative models, the discussion about digital rights management is more intense than ever before. Digital content is coming up more and more on the Internet, and doubts that technical protection alone will suffice are increasing. Existing technical approaches to digital rights protection are proprietary and mutually incompatible, they establish closed user groups. Moreover, they overestimate the technical possibilities of transferring usage rules of physical goods to the digital world by means of control of end-user devices. When technology lacks, law will called in to settle the dispute.

Online Marketplace

Selling the 'Identity' of online games can be financially lucrative when the identity's reputation is highly rated or well-respected. In most role-playing games, players can swap items within the game using the game's virtual currency. But many players prefer to earn real money, selling items and characters on auction sites such as eBay or specialty barter sites. A random search of eBay showed more than 150 items, including online accounts with several highly-developed characters selling for $300 or more. Although in January 2007 eBay discontinued the selling of virtual artefacts, several

---

[322] J Dibbell, 'A Rape in Cyberspace, or How an Evil Clown, a Haitian Trickster Spirit, Two Wizards, and a Cast of Dozens Turned a Database into a Society' in: *The Village Voice* (December 21, 1993) pp. 36 42. WWW: <http://www.ludd.luth.se/mud/aber/articles/village_voice.html> and WWW: <http://www.juliandibbell.com/texts/bungle.html>.

[323] Digizerk. Source: WWW: <http://www.steenhouwerij-rijtink.nl/download/persbericht_eng.pdf>.

alternatives quickly emerged. In the wake of eBay's decision to halt auctions of virtual property, new companies entered the market to fill the void, including one that allows gamers to trade game currency directly with one another rather than to buy it from exchanges. [324] One company, Sparter, says (in 2007) this eBay-like 'peer-to-peer' approach will result in lower prices as sellers compete. It incorporates a reputation system and escrow for gold delivery.

Economic Emerge of New Services

New services result in new discussions about identities. In situations of (varieties of) Outsourcing, Remote Computing Services, Facilities Management, user identities of licences and facilities are transferred 'on behalf of' the regular user/owner. *Web 2.0* is a 'social phenomenon referring to an approach to creating and distributing Web content itself, characterised by open communication, decentralisation of authority, with freedom to share and re-use, and 'the market such as a conversation.' [325] E.g.:
- *Blogging:* This type of collaborative Website is operated by one person, but (in chronological order) entries can be added by any other (mostly anonymous) user(s).
- *Blook:* Like blogging, the story like a book is added by many other (mostly anonymous) user(s).
- *Wikipedia:* An online encyclopaedia based on the notion that an entry can be created by any user and edited by any other user. IPs of editors are logged constantly.
- *Flickr:* A online photo book based on the notion that each of us share our photos.
- *Social networking:* Through the marketplace the members of the social networking website can make full use of the dynamic space by exchanging products, knowledge and services.

With novelties such as Open Source, Open Data, and Open Content as well, Web 2.0 is an attitude not a technology. It's about enabling and encouraging participation through open applications and services. One of the propagated characteristics of Web 2.0 is collaboration between Websites and events (generated by 'things') from users in order to create and enhance the content of the database.

Online Tax

The US Congress announced in 2006 it would investigate the amount of commerce taking place in virtual game worlds. [326] It said it was interested solely in the 'universe of transactions' that occurs within online worlds such as SecondLife. Although an economic value can be put on this trade, as in-game currencies do have an equivalent real world value, the investigation was not being carried out with a view to slapping taxes on this trade.
Shortly after the Australian Tax Office warned citizens to consider whether their gaming was 'a hobby or a business', otherwise they would have to pay income tax on their real profit of virtual property. [327] Virtual lifers making virtual fortunes in virtual worlds such as 'World of Warcraft' or 'SecondLife' could face real tax bills.

Users of Virtual Identity

Human
Individuals, groups and organisations use virtual identities for various purposes – as explained earlier – but they are always as a kind of electronic representative in computers, systems, and/or networks. In most applications they control their virtual identity themselves. Sometimes the 'clone' is acting autonomously 'such as the original' within a small pre-programmed field.

Machine
Bots, chatterbots, chat bots, web bots, avatars, NPC's or virtual friends, they all are a kind of virtual software robots with a pre-programmed, automated identity. Some call them Automated Service Agent™ or Verbots® (verbally-enhanced software robots), others call them Robot Agents. They are

---

[324] Peer-to-Peer Market for RMT Trading Debuts in: 'Virtual Worlds, Real Profit. Real Money Trading in MMORPG's'. WWW: <http://www.virtualeconomies.net/blog/2007/02/14/ new-peer-to-peer-system-for-rmt-trading-debuts/>.

[325] M Sigal, '*What the Hell is Web 2.0*? The Great Web Mash-up Begins', (Sept. 16, 2005). WWW: <http//www.oreilleynet.com/pub/wlg/7825>.

[326] US Joint Economic Committee: 'Virtual economies need clarification, not more taxes' Press Release #109-98, 17-10-2006. WWW: <http//www.house.gov/jec/news/news2006/pr109-98.pdf>.

[327] NN, 'Virtual world: tax man cometh' (2006). WWW: <http://www.theage.com.au:80/news/ biztech/virtual-world-tax-man-cometh/2006/10/30/1162056925483.html>.

all a combination of some Artificial Intelligence, Natural Language Processing, and creativity. More and more they are able to converse in natural language with humans, with minimal frustration and increasing satisfaction. People ask questions in their own words and get direct answers, not pages of search results.

> *'Running free in cybernetic spaces, proliferating and transmuting at unpredictable velocities, autonomous digital entities operate at a further remove from the gaze or touch of the human agent than any prior generation of forms'*
>
> Clark[328]

However you refer to them, bots allow you to create an engaging virtual personality that can perform many tasks - from assisting with common computer tasks, to serving as a virtual assistant that can help you get organised, to acting as a teacher that can play games or administer tests. A bot can even be a politician in the 2006 Dutch party campaign, answering questions about the party manifesto.[329] Whatever you do with your bot, the possibilities are endless.

> *'Cupid has a total of ten different avatars. He changes his look mostly with imaginary parameters of romance, love, flirting and sex. Small visual tricks add a lot to his characteristics. Some of the keywords and cupid's lines are designed to some particular avatar, which he keeps wearing until another keywords triggers a new look. Parts of his dialogue he just says with the avatar he happens to be wearing, which adds a level of different possible meanings with the element of randomness and surprise: the same line with a different avatar might create a different meaning.'*
>
> Leena Saarinen in Bots' Avatars [330]

New guidelines will be drawn worldwide to ensure human control over robots, protecting data acquired by robots and preventing illegal use. These key considerations could reflect the three laws of robotics put forward by author Isaac Asimov in his short 1942 story 'Runabout' [331].
The European Robotics Research Network (EURoN.org) presented March 2006 a set of guidelines on the use of robots. [332] It stated: 'In the 21st Century humanity will coexist with the first alien intelligence we have ever come into contact with - robots.' This ethical roadmap has been assembled by researchers who believe that robotics will soon come under the same scrutiny as disciplines such as nuclear physics and bioengineering. In 2006 a UK government study predicted that in the next 50 years robots could demand the same rights as human beings. [333]
An ethical code to prevent humans abusing robots, and vice versa, is being drawn up by South Korea. The Robot Ethics Charter will cover standards for users and manufacturers and will be released in 2008. 'Imagine if some people treat androids as if the machines were their wives. Others may get addicted to interacting with them just as many Internet users get hooked to the cyberworld.' [334]

Avatar Rights

Lastowka and Hunter discussed whether avatars have enforceable legal and moral rights. [335] Avatars, the user-controlled entities that interact with virtual worlds, are a persistent extension of their human users, and users identify with them so closely that the human-avatar being can be

---

[328] N Clark, *Rear-View Mirrorshades: The Recursive Generation of the Cyberbody*, in: M Featherstone, and R Burrows (eds), 'Cyberspace/Cyberbodies/Cyberpunk : Cultures of Technological Embodiment.' (Sage, London 1995) p 130.

[329] Christian Democratic Appeal (14-11-2006) (content changed a little after 1st access; 2nd access: 01-07-2007). WWW: <http://verkiezingen.cda.nl/index.asp?id=73&utm_id=BAN0100&utm_source=msn>.

[330] L Saarinen, 'Chatterbots: Crash Test Dummies of Communication' (2001); WWW: <http://mlab.uiah.fi/~lsaarine/bots/chapter3.html>.

[331] I Asimov, *Runabout,* (NY 1942). I Asimov: *I, Robot.* (Doubleday & Company, NY 1950).

Since 1726 science fiction authors present a future where only behavioural restrictions on robots stand between peace and destruction. See: HB Franklin, 'Computers in Fiction' (2000); WWW: <http://newark.rutgers.edu/~hbf/compulit.htm>. Such restrictions, however, are unethical because they violate the robot's free-wills. Rather than content-based restrictions on free-will, robots need mental structures that will guide them towards the self-invention of good, ethical behaviours.

Isaac Asimov defined the three laws of robots:
1. A robot may not injure a human being or, through inaction, allow a human being to come to harm.
2. A robot must obey orders given it by human beings except where such orders would conflict with the 1st Law.
3. A robot must protect its own existence as long as such protection does not conflict with the 1st or 2nd Law.

[332] WWW: <http://www.roboethics.org/site/modules/mydownloads/download/ROBOETHICS%20ROADMAP%20Rel2.1.1.pdf>.

[333] BBC: 'Robots could demand legal rights', 21-12-2006. WWW: <http://news.bbc.co.uk/1/hi/technology/6200005.stm>.
Cited: 'The Sigma and Delta Scans', UK Office of Science and Innovation's Horizon Scanning Centre. WWW: <http://www.foresight.gov.uk/HORIZON_SCANNING_CENTRE/ Strategic_Horizon_Scans/ Strategic_Horizon_Scans.html>.

[334] BBC: 'Robotic age poses ethical dilemma', 07-03-2007. Ref. to Three Laws of Robotics (plus 1 attached). WWW: <http://news.bbc.co.uk/2/hi/technology/6425927.stm>. `Robot Ethics Charter` to be made.

[335] FG Lastowka and D Hunter, 'The Laws of the Virtual Worlds'. *California Law Review* (2003); WWW: <http://a.parsons.edu/~imagined_realms/lastowka.hunter.lawsofvirtualworlds.pdf>.

thought of as a cyborg. If the avatar of an intelligent agent is granted legal personality, should it have publishing rights of the creators of personal likenesses? The issue of cyborg rights within virtual worlds, and whether they may have real world significance, is important too.

Effects of the Use of Virtual Identity

The use of virtual identities is increasing, both by the ongoing use in identity-related software applications, and by the ways of representing a managed identity in Internet contacts. When a person abstracts himself from the roles of a good citizen, a good spouse, and the good friend (s)he is playing, this person creates virtual identities as a result of the effects on some inner reactions. This virtual identity especially finds a way to express itself by existing on the Internet, transferring feelings and thoughts. Like the effect of watching television, there is a relation between increased Internet usage and increased psychological depression. Therefore, some researches say that the usage of Internet may negatively affect the social relations in society.

Research on the impact of Internet use on social ties has generated conflicting results. Based on data from the 2000 General Social Survey, another study finds that different types of Internet usage are differentially related to social connectivity. While nonsocial users of the Internet do not differ significantly from nonusers in network size, social users of the Internet have more social ties than nonusers do. Among social users, heavy email users have more social ties than do light email users. There is indication that, while email users communicate online with people whom they also contact offline, chat users maintain some of their social ties exclusively online. These findings call for differentiated analyses of Internet uses and their effects on interpersonal connectivity. [336]
Other recent research (e.g. at Secondlife, see p 214) show that the social effects are negligible when the Internet usage is regular and not addictive.

Due to the complexity of the online behaviour (including transactions) management of the virtual identity is required. Rutter & Smith researched the management of identity in the everyday newsgroup interaction. [337] The sense of community that exists in the newsgroup relies heavily on the posters' ability to know with whom they are interacting. A practised familiarity with others allows members to understand the nature of their online relationships, assess the validity of information offered to them by others, and place in context comments and actions of other posters. Unlike the often-fantastical environments of some synchronous online interaction, the identities enacted in the newsgroup are taken to be 'real' in a serious sense. When messages are posted to the group or address individuals, a level of trust is offered and expected between those involved in the group. The conclusion is that a well-managed virtual identity can support to the trust in a virtual social community.

Summary Chapter 5 'Forms, Varieties and Valuables of VIDs'

After the coming into being by virtual reality is clarified in the first chapters, the virtual world is pointed out in this chapter as an environment that enables extraterrestrial and superterrestrial experiences. The forms and varieties of virtual identities are explored, and the use and users indicated. Finally a list of examples of the broad use of virtual identities, both user (human) and machine generated.
The key feature of cyberspace is the interaction through which a new sense of self and control can be constructed. The result of these new senses of self is a new sense of presence that fills the space with fluid forms of network and community. The basis of the community of people interacting in a technological environment is shifting from culture-defining mass media to a proliferation of media as alternative sources of mediated experience. People are sufficiently concerned about their digital identity to take steps to protect it and already an appeal has been made on programmers to embed privacy capabilities in software, which implicitly means that virtual identities are taking the main role in the software development. In the next chapter the anonymity of virtual identities will be examined.

---

[336] S Zhao, 'Do Internet users have more social ties? A call for differentiated analyses of Internet use'. *Journal of Computer-Mediated Communication, 11*(3) (2006), article 8.

[337] J Rutter and GWH Smith, 'Presenting the offline self in an everyday online environment.' Paper presented at the ' Identities in Action', Wales (1995). WWW: <http://www.cric.ac.uk/cric/staff/Jason_Rutter/papers/Self.pdf>.

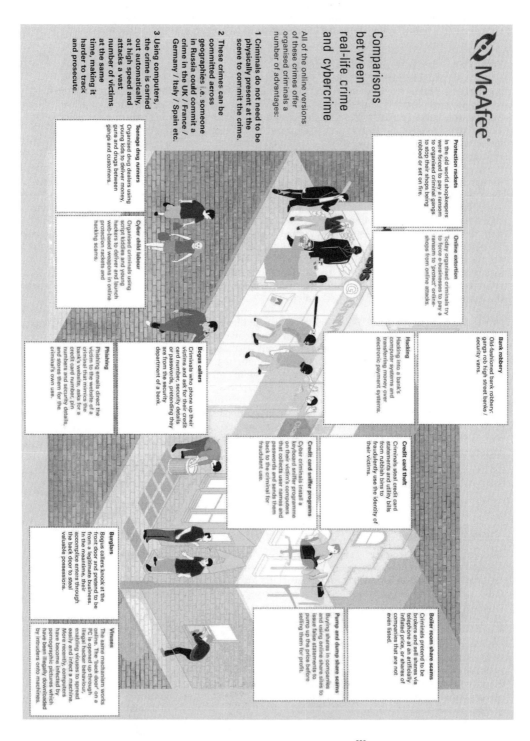

Figure 33 - Comparisons between real-life crime and cybercrime (© McAfee) [338]   See more at page 151.

---

[338] Source: WWW: <http://www.trustmarquesolutions.com/security/documents/mcafee_crime_cybercrime.jpeg>.

6 Virtual Identity and Anonymity

'If all cyberspace gives you is an email address, a place to hang your virtual hat and chat about your hobbies, you've been cheated. What most of us want is a place where we're known and accepted on the basis of what Martin Luther King, Jr. called 'the content of our character.'

Mike Godwin [339]

Introduction
Almost 2.500 years ago people used pseudonyms to be anonymous, and imaginary worlds as well as the jurisdiction of law were discussed. For over than 150 years there was a de facto Internet, with chat rooms, virtual affairs and online identities. Going back in history, discussions were also held about the shameless youth that handily used technology, about rude colonists who made their own rules in the conquered territories, and about floating people who were hopping between imaginary and realistic societies. This chapter discusses the importance of pseudonymity and anonymity for the development of virtual identity. The challenge and value of VID for individual and society are determined. The anonymous 'business as usual' is additionally viewed.

Anonymity and Pseudonymity
The motto 'to know you is to love you' [340] seems to be the driver for commercial engagement in the Internet identity policy. This subchapter points out the acceptance of certain anonymity as long as the business gets grip of the person behind the mask. It focuses on the discussion about a certain policy regarding anonymous communication on the Internet and the way virtual identity can influence that policy.

Anonymity, Technology and Virtual Identity
Virtual Identity can be anonymous but is not equal to anonymity. Anonymity is derived from the Greek word ανωνυμία, meaning 'without a name' or 'nameless', and more originally mean 'without a law'. In colloquial use, the term typically refers to a person, and often means that the personal identity or personally identifiable information of that person is not known. More strictly, and in reference to an arbitrary element (e.g. a human, an object, a computer), within a well-defined set (called the 'anonymity set'), the 'anonymity' of that element refers to the property of that element of not being identifiable within this set. If it is not identifiable, then the element is said to be 'anonymous'.

Whistle-blowers usually want to stay anonymous when they report about potential negligence, nevertheless many companies and governments avoid anonymous reporting. The French Data Protection Act 1978 (amended 2004) declares that the threat to a person's privacy and reputation through potential malevolent and anonymous reporting could be of greater harm than the damage the whistle-blowing scheme aims to prevent. [341]
The suggested solution, therefore, is that whistle-blowing schemes should operate in such a way so as not to encourage anonymous reporting as the prescribed way to make a complaint. In particular, companies should not even advertise the fact that anonymous reports may be made through a scheme. [342]

---

[339] M Godwin, 'Nine Principles for Making Virtual Communities Work' *Wired;* Issue 2.06 - Jun 1994. p72-73. WWW: <http://www.wired.com/wired/archive/2.06/vc.principles.html>.

[340] Peter & Gordon's 1965' popular record 'To Know You Is To Love You', written by Phil Spector, was a cover version of The Teddy Bears' #1 hit from 1958, 'To Know Him Is to Love Him.'

[341] Chapter XI, Article 67. Act n°78-17 of 6 January 1978 on 'Data Processing, Data Files and Individual Liberties' (Amended by the Act of 6 August 2004 relating to the protection of individuals with regard to the processing of personal data); WWW: <http://www.cnil.fr/fileadmin/documents/uk/78-17VA.pdf > and WWW: <http://www.cnil.fr/index.php?id=4>.

[342] In the 2006 ed. of the study 'Data Protection in the European Union and other Selected Countries', a report prepared jointly by law firms working together on a best friend basis, is it argued that failures regarding privacy compliance also may result in damages to the organisation's reputation, brand or business relationships as well as a legal liability or regulatory sanctions, lawsuits for deceptive business practices, and loss of customers' and employees' confidence. Compliant privacy practices are a key element of corporate governance and accountability. WWW: <http://www.ey.com/global/download.nsf/Italy/ Data_Protection_in_the_European_Union_and_other_ Selected_Countries/$file/10541%20Data%20Protection%20Italy.pdf>.

Such a scheme informs the whistle-blower, at the time of establishing the first contact with the scheme, that their identity will be kept confidential at all stages of the process (and is changed in an anonymous virtual identity). If, despite this information, the person reporting to the scheme still wants to remain anonymous, the report will then be accepted into the scheme.

Sometimes it is desired that a person can establish a long-term relationship (such as a reputation) with some other entity, without their personal identity being disclosed to that entity. In this case, it is useful for the person to establish a unique identifier, called a pseudonym, with the other entity. In the digital environment of information technology a pseudonym will at least be an alias but it also could be a virtual identity.

> A pseudonym enables the other entity to link different messages from the same person and, thereby, the maintenance of a long-term relationship. Although typically pseudonyms do not contain personally identifying information, communication that is based on pseudonyms is often not classified as 'anonymous', but as 'pseudonymous' instead. Indeed, in some contexts, anonymity and pseudonymity are separate concepts. However, in other contexts what matters is that both anonymity and pseudonymity are concepts that are, among other things, concerned with hiding a person's legal identity. In such contexts people may not distinguish between anonymity and pseudonymity. [343]

Examples of pseudonyms are user names, nicknames, credit card numbers, student numbers, bank account numbers, IP network addresses.

When communicating with others over the Internet most people prefer to not use any kind of identifiable handle, such as a user name, nor any arbitrary way of identifying who is communicating. Even one user name can be used by more people.

- *Example 1*: At the Japanese forum 2channel [344] the famous poster with the user name 'anonymous' appears to be one person, which is obviously impossible given the number of posts attributed to that name.
- *Example 2*: In the Netherlands, in 1993, a male journalist created an anonymous identity 'Truus' as a 'web bitch' that kept the discussion going until 1997. [345] Other people were permitted to use this ID.

In past there have been significant anonymous postings, before the cybernetic period too:
- *Example 3*: In 1787 a series of articles under the pseudonym '*Publius*' appeared in a number of US newspapers, primarily those of New York [346] but soon became pervasive, followed by quite a lot of literature on this type of authorship. [347]

When sending messages over the Internet, many people enjoy a sense of anonymity (or at least pseudonymity). Many popular systems, such as email, instant messaging, MSN, web forums, market places, Usenet, and Peer-to-Peer (P2P) systems, foster this perception because there is often no obvious way for a casual user to connect to other users with a 'real world' identity.

Anonymous Identity in the Information Age

In 1999 an 'Ethical and Legal Aspects of Human Subjects Research on the Internet' workshop was held.[348] When discussing the topic '*The Meaning of Anonymity in an Information Age*,' the philosopher Helen Nissenbaum asked the audience 'What is anonymity? And, what are we seeking to protect when we propose to protect it?'

---

[343] Anonymity; WWW: <http://en.wikipedia.org/wiki/Anonymity>.

A pseudonym or pen name may be used by an author of a copyrighted work. A work is pseudonymous if the author is identified on copies or phonorecords of that work by a fictitious name (nicknames or other diminutive forms of one's legal name are not considered "fictitious"). As is the case with other names, the pseudonym itself is not protected by copyright. WWW: <http://www.copyright.gov/fls/fl101.pdf>.

[344] 2channel (2 ちゃんねる ni channeru) is thought to be the largest Internet forum in the world. With over 10 million visitors every day (as of 2001), it is gaining significant influence in Japanese society, approaching that of traditional mass media such as television, radio, magazines. WWW: <http://2chan.net>.

[345] Miss Truus de Wit was an anonymous phenomenon in the Dutch part Usenet; WWW: <http://www.truus.com/>.

[346] C Lucock, 'Blog*on*nymity - blogging On the Identity Trail' PUBLIUS, *The pseudonym and poetry* (2006); WWW: <http://www.anonequity.org/weblog/archives/2006/09/publius_the_pseudonym_and_ poet.php>.

[347] MH Spencer, 'Anonymous Internet Communication and the First Amendment: A Crack in the Dam of National Sovereignty'. In: *3 VA. J.L. & TECH.* 1 (Spring 1998). WWW: <http://www.vjolt.net/vol3/issue/vol3_art1.html>.

[348] Workshop convened by the AAAS Program on Scientific Freedom, Responsibility and Law in June 1999. WWW: <http://www.aaas.org/spp/sfrl/projects/intres/report.pdf>.

> 'The natural meaning of anonymity can be defined as the natural meaning of anonymity, as may be reflected in ordinary usage or a dictionary definition, is of remaining nameless, that is to say, conducting oneself without revealing one's name. Extending this understanding into the electronic sphere, one might suggest that conducting one's affairs, communicating, or engaging in transactions anonymously in the electronic sphere, is to do so without one's name being known. Specific cases that are regularly discussed include:
>
> - sending electronic mail to an individual, or bulletin board, without one's given name appearing in any part of the header
> - participating in a 'chat' group, electronic forum, or game without one's given name being known by other participants
> - buying something with the digital equivalent of cash
> - being able to visit any Website without having to divulge one's identity
>
> The concern I wish to raise here is that in a computerised world concealing or withholding names is no longer adequate, because although it preserves a traditional understanding of anonymity, it fails to preserve what is at stake in protecting anonymity. Why? Information technology has made it possible to track people in historically unprecedented ways. We are targets of surveillance at just about every turn of our lives. In transactions with retailers, mail order companies, medical care givers, day-care providers, and even beauty parlours, information about us is collected, stored, analysed and sometimes shared. Our presence on the planet, our notable features and momentous milestones are dutifully recorded by agencies of federal, state and local government including birth, marriage, divorce, property ownership, drivers licenses, vehicle registration, moving violations, passage through computerised toll roads and bridges, parenthood, and, finally, our demise. Into the great store of information, we are identified through name, street address, email address, phone number, credit card numbers, social security number, passport number, level of education and more; we are described by age, hair color, eye color, height, quality of vision, purchases, credit card activity, travel, employment and rental history, real estate transactions, change of address, ages and numbers of children, and magazine subscriptions. The dimensions are endless.' Nissenbaum [349]

On the Internet the acceptance of anonymity is differentiated by contexts, such as electronic forums that have different properties and rules regarding anonymous communication. For example, an adult chat room may allow people to use pseudonyms while a children's room may not.

In 1999 Gary Marx suggested seven ways of identifying a person, including their legal name, their location, and distinctive appearance and behavioural patterns. [350] Marx makes the important point that identity and anonymity are features of social relationships rather than a property of a person. (Peter is not anonymous, although he may communicate anonymously or pseudonymously with Susan). He identified some major rationales and contexts for anonymous communication (free flow of communication, protection, experimentation) and identifiability (e.g., accountability, reciprocity, eligibility). Marx's principle of truth in the nature of naming holds that those who use pseudonyms on the Internet in personal communications have an obligation to indicate they are doing so. This principle provoked intense and sustained discussion at the conference. Lastly, Marx suggests 13 procedural questions to guide the development and assessment of any policy regarding anonymous communication on the Internet.

In the same period, Froomkin examined how this controversy will have direct effects on the freedom of speech, the nature of electronic commerce, and the capabilities of law enforcement. [351] He showed how the legal resolution of the anonymity issue is also closely bound up with other difficult and important legal issues: campaign finance laws, economic regulation, freedom of speech on the Internet, the protection of intellectual property, and general approaches to privacy and data protection law. He argued that the legal constraints on anonymous communication and the constitutional constraints on those who would regulate it further should be considered in tandem with the policies driving regulation as well as the side effects. The regulation of anonymous and pseudonymous communications promises to be one of the most important and contentious Internet-related issues of the next decade.

---

[349] H Nissenbaum, 'The Meaning of Anonymity in an Information Age' The Information Society, 15, pp 141-144 (1999) (Reprinted in Readings in CyberEthics 2001). WWW: <http://www.nyu.edu/projects/nissenbaum/paper_anonimity.html>.

[350] GT Marx, 'What's in a Name? Some Reflections on the Sociology of Anonymity' The Information Society 15/2 (April-June 1999); WWW: <http://web.mit.edu/gtmarx/www/anon.html>.

[351] M Froomkin, 'Legal Issues in Anonymity and Pseudonymity' AAAS Symposium Vol.15, The Information Society (1999) p 113.
M Froomkin, 'The Death of Privacy?' Stanford Law Review, Vol. 52 (2000) p 1461. WWW:
<http://cyber.law.harvard.edu/privacy/Fromkin_DeathOfPrivacy.pdf> and WWW: <http://www.law.tm/>.

Online Anonymity

One of the most significant and for users attractive features of the electronic systems and networks is the seeming anonymity for communications and (trans-)actions.[352]
Because of this they are becoming annoying and sometimes start cyber-pestering.[353]

It is stated also that the first reason to buy in e-market shops appears not to be low prices or convenience, but being invisible as buyer or seller.[354] However, in practise web shops follow your click behaviour, and e-markets such as Amazon know more about your browsing track record than you remember yourself. Although Danes do not have very much faith in consumer protection in other EU countries, the study shows that many consumers do not protect themselves, or do not know how to protect themselves when they shop on the Internet. Among people who shopped on the Internet in another European country within the past year. (figure 34) [355]

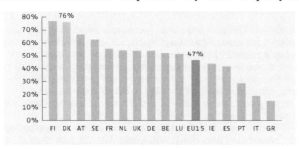

Danes have a high degree of trust that the Danish consumer protection legislation is complied with. Only the Finns have a stronger belief that their consumer protection legislation is complied with.

Source: Special Eurobarometer (2003), Consumer protection in the EU, p. 19.
Figure 34

Using an online identity (alias) as a (second) social identity lets Internet users establish themselves in online communities and web shops. Although some people use their real names online, most Internet users prefer to identify themselves by means of pseudonyms, which reveal varying amounts of personally identifiable information.

In some online contexts, including Internet forums, Multi User Dimension (MUDs), instant messaging, and massively multiplayer online games (MMOGs), users can represent themselves visually by choosing an avatar, an iconised graphic image. Some others use a music or text fragment to distinguish themselves. As other users interact with an established online identity, it acquires a reputation, which enables them to decide whether the identity is trustworthy.

Opponents of online anonymity suggest that anonymity enables or encourages illegal or dangerous activity (terrorism, drug trafficking, sexual advances towards minors, and so on). The inference of such claims is that, without anonymity, these things would not occur or be less likely to occur to the same extent, and members of society would be safer. This is simply countered by pointing out that prior to the practical availability of this kind of anonymity, all these things occurred anyway.

Capurro and Pingel point out that a solely metaphysical distinction between bodily and digital identities would blur the richer view of existential identity, e.g. if a research report fails to anonymise the pseudonyms used in a chat room by specific persons. [356] A simple metaphysical dichotomy between offline and online identity may lead to unethical consequences when the multiple ways are ignored in which embodied persons are connected with and emotionally invested in their online identities as part of their existential choices and projects. Very real harm can follow when linked information about online identities is revealed.

The limits of a metaphysical distinction are clear when we consider the history of legal protection of personal data - for instance in Germany - and the problems faced by such legislation when related to digitised data collected by different private and/or political bodies for different purposes, concluded Capurro and Pingel.

---

[352] NN, 'Many Net users prefer online anonymity, poll finds.' Seattle Times, June 16, 1996.

[353] J Noveck, 'Online Anonymity Lets Users Get Nasty'. March 21, 2007 11:21AM; WWW: <http://www.newsfactor.com/news/Online-Anonymity-Lets-Users-Get-Nasty/story.xhtml?story_id=131000F9XDED>.

[354] Danish consumers in the European e-market; WWW: <http://www.forbrugereuropa.dk/fileadmin/Filer/ Forbrugereuropa_-_pdf/Danish_consumers_in_the_European_e-market.pdf>.

[355] Source: EU Special Eurobarometer. Issues relating to business and consumer e-commerce. 2003/ Publ. 2004. WWW: <http://ec.europa.eu/public_opinion/archives/ebs/ebs252_en.pdf>.

[356] R Capurro and C Pingel, *Ethical Issues of Online Communication Research* (2000); WWW: <http://www.nyu.edu/projects/nissenbaum/ethics_cap_full.html>.

Combinations of digitised online data can be a threat to privacy that allows not only governments (and criminals!) but also private entrepreneurs to have a detailed view of individuals and groups, their desires, interests, occupations, etc., thus giving rise to much more comprehensive kinds of manipulation and control than with conventional (particularly paper-based) media.

Personal Profile in both the Real and Virtual Environments

The digital identity is a model based on information about our existence in the online world and it's used as replacement of our real world identity. Based on this digital identity a person will be judged or decisions will be made for or about him. This differs from the real world, where personal identity results from human interaction.

In his theory about personal identity Goffman states that personal identity comes into being by social interaction. [357] We socialise with others and in that context we send out specific signals that express how we want others to see us. We also unwittingly send out signals to others. All these signals represent an image of who we really are. Other people react to that presentation. Personal identity is partly shaped by oneself and partly by the interaction with others. In face-to-face contact facts such as gender, race, clothes and non-verbal aspects play an important role. Ruesch and Bateson point out that in communication these aspects determine how others see us. [358]

Davis discusses in 'Identity Ambivalence' several intriguing theories about the social and psychological significance of fashion in modern culture. [359] What makes clothes fashion; how fashion choices express social status, gender identity, sexuality, and conformity; and how fashion is (or is not) accepted are all discussed. In various ways fashion seeks constantly to get those attuned to its symbolic movements to alter their virtual identities (Goffman)[360], to relinquish one image of self in favour of another, to cause what was until then thought ugly to be seen as beautiful and vice versa.

To Be or Not to Be Identified

This subtitle concerns the field of tension between commerce, state and individual related to the grade of anonymity of a person's identity on the Internet. The value and importance of virtual identity for the society is discussed, as well as the reasons for the business partners to require authenticity before entering the World Wide Web.

Political View to Identity @ The Internet

Privacy and Identity

Politicians and their staff have had and still have an exceptional opinion and a biased view of the Internet phenomenon. For years they judged the Internet as a playground for scientists, kids, and nerds, something that will pass by when the game was over. 'The Internet - as trend watched in 1997' made them wonder about the 'thousands of such newsgroups, each serving to foster an exchange of information or opinion on a particular topic running the gamut from, say, the music of Wagner to Balkan politics to AIDS prevention to the Chicago Bulls.' [361]
Some years later Web 2.0 was experienced as a commercial buzzword to link people when Information Technology and the Internet were only 'interesting topics' for the political sciences, not something for the political agenda. [362] So, for years there has been no concrete policy for the Internet, not even on a global level.

---

[357] E Goffman, *The presentation of self in everyday life* (Doubleday, Garden City NY 1959).

[358] J Ruesch and G Bateson, *Communication: The Social Matrix of Psychiatry*, [1951](1968) p 169.
Gregory Bateson's contributions influenced the early days of cybernetics, during the times of the Macy Foundation conferences, when priorities still centered on designing systems that could use their own output as input (such as guiding systems for ground to air missiles-a remnant memory from World War II). It offered alternatives to cause and effect explanations. At the Macy conference in 1942, Bateson met Warren McCulloch and Julian Bigelow who were talking about a new concept, 'feedback' which Bateson felt he lacked when writing 'Naven', one of his longer works, in 1936. At this time he also met Norbert Wiener, John von Neumann and Evelyn Hutchinson, all of whom have influenced his thinking and writing since 1942.

[359] F Davis, *Fashion, Culture, and Identity*. (University Of Chicago Press 1994).

[360] E Goffman, *Stigma: Notes on the management of spoiled identity*. Prentice-Hall, Englewood Cliffs, NJ: 1993).

[361] The citation is a description of the Internet, set forth by Justice John Paul Stevens in the landmark Reno v. ACLU decision, June 26, 1997. WWW: <http://www.gseis.ucla.edu/iclp/internet.html>.

[362] Keele Guide to Political Science on the Internet. WWW: <http://www.keele.ac.uk/depts/por/psbase.htm>.

The EU initiated the Roadmap for Advanced research in Privacy and IDentity management (RAPID) [363] followed by the PRIME (Privacy and Identity Management for Europe) project, sponsored by the European Commission and the Swiss Government, aiming to restore the dignity of an individual's privacy sphere in an increasingly online world. The objective of this project is to give 'individuals sovereignty over their personal data', and to enable 'individuals to negotiate with service providers the disclosure of personal data and conditions defined by their preferences and privacy policy'. This project calls into play some fundamental principles: user support, openness, consent, accuracy and completeness, data minimisation, notification, security, and access to law enforcement.

In the Far East *ad hoc* research has started on policy matters affecting Internet networking in the Asia-Pacific region. [364]
Africa and South America don't show progress yet.

The events of 11 September 2001 caused the US government to rush many methods of investigation into law and now it faces serious conflicts when applying methods of digital surveillance that may interfere, for better or for worse, with the bodily and digital life projects of people.
Following these digital surveillance manoeuvres by the US government in 2001-2002, some other governments argued that the policy behind the technical functioning of the Internet is not in line with the principle of sovereign equality that regulates the relationship among nations. This principle lies down in the Charter of the United Nations as *a jus cogens* norm of the contemporary international law.
Brazil, China, India, South Africa and other governments called for a greater role in the management of the global Internet core. They argue that these core resources are crucial, not only for their region but also for the functioning of the national Internet infrastructure and insofar part of their national interests. These governments acknowledged the driving role of the US government in the early days of the Internet, but made it clear that 'what was good of an Internet with one million users would be not good enough for an Internet with one billion users'.

The United Nations World Summit on the Information Society (WSIS) started a debate on 'Internet governance' in 2002. So, finally the United Nations turn their attention to the government of the Internet, but still without taking decisive measures to counter Internet governance. [365] An ICT Taskforce was harnessed and in 2005 the Working Group on Internet Governance launched 'The Case for Establishing a Global Internet Council'. Al-Darrab argues that 'the involvement of States in the overall Internet governance process through a Global Internet Council would provide international legitimacy and ensure accountability for ICANN and other existing and future Internet governance institutions where required. It would also legitimise the governance process in the eyes of national Governments and facilitate agreements on Internet-related public policies, as well as provide international legitimacy to dispute resolution and arbitration procedures relating to international intellectual property rights. The Council also would facilitate full participation in Internet governance arrangements by developing countries.' [366]

Meanwhile, the International Telecommunication Union (ITU) – the UN's specialised agency for telecommunications since 1865 – examined in 'The Digital Life' [367] the rapidly changing technological and social environment surrounding the individual (later referred to as the 'digital individual') and the blurring boundaries between the public and private spheres of existence.

The Digital Individual. From Person to Personae.

'The complexity of the interaction between technology, personal consumption and the construction of virtual identity in cyberspace has traditionally been ignored, but is now the subject of observation in many quarters. Users of digital technologies today have a wide scope for constructing their identity. The mostly nameless and faceless environments of cyberspace create an ideal background for developing alternate identities or digital personae. Unlike face-to-face interaction, it is much more difficult to categorise people online according to age, gender, race, country of residence, social class, body shape etc. Consequently, users may feel more inclined to interact in what seems to them a more

---

[363] RAPID (Roadmap for Advanced research in Privacy and IDentity management) EU project IST-2001-38310 (2003). WWW: <http://europa.eu.int/information_society/activities/egovernment_research/doc/rapid_roadmap.doc>.

[364] Connecting People - Changing Lives in Asia (2005). WWW: <http://www.idrc.ca/en/ev-94703-201-1-DO_TOPIC.html>.

[365] ICT Policy and Governance Working Group; WWW: <http://www.unicttaskforce.org>. Report of Working Group on Internet Governance (2005); WWW: <http://www.wgig.org/docs/WGIGREPORT.doc>.

[366] A Al-Darrab, *The need for international internet governance oversight*; in: WJ Drake (2005): 'Reforming Internet Governance: Perspectives from WGIG'. WWW: <http://www.wgig.org/docs/book/WGIG_book.pdf>.

[367] ITU, *The digital life*. Internet Report 8 (2006); WWW: <http://www.itu.int/osg/spu/publications/digitalife/docs/digital-life-web.pdf>.

> *anonymous and forgiving world. Moreover, the Internet makes it fairly easy for individuals to create multiple representations of their identities, mainly due to the lack of a generic system for identification. This fragmentation of identities can be accidental, but also intentional. Creating more than one identity can even be desirable to some, depending on the context and exchanges involved. For instance, a user may wish to be aggressive and egotistical in one context (e.g. in a multiplayer game), but sensitive and sociable for virtual encounters of the romantic kind.'*   The Digital Life Report

The Next Mobile Revolution

In 2001 Hans Snook, CEO of Telecom Orange, described the mobile phone as a 'remote control for life. [368] The mobile phone has developed from voice to text to pictures to data processor to a sort of bionic buddy. [369] [370] It has become such an intimate and important aspect of a user's daily life that it has moved from being a mere technical tool to an indispensable social escort. [371] Its highly personalised and emotive nature has meant that its form and use have begun to represent the very personality and individuality of its user. In other words, it has in some respects become a reflection of a user's identity.

> *'... The next techno-cultural shift-ashift will be as dramatic as the widespread adoption of the PC in the 1980s and the Internet in the 1990s. The coming wave is the result of super-efficient mobile communications-cellular phones, personal digital assistants, and wireless-paging and Internet-access devices that will allow us to connect with anyone, anywhere, anytime. From the amusing ('Lovegetty' devices in Japan that light up when a person with the right date-potential characteristics appears in the vicinity) to the extraordinary (the overthrow of a repressive regime in the Philippines by political activists who mobilized by forwarding text messages via cell phones). Examples of the fundamentally new ways in which people are already engaging in group or collective action. ... Consider the dark side of this phenomenon, such as the coordination of terrorist cells, threats to privacy, and the ability to incite violent behavior. This is penetrating perspective on the brave new convergence of pop culture, cutting-edge technology, and social activism. The real impact of mobile communications will come not from the technology itself but from how people use it, resist it, adapt to it, and ultimately use it to transform themselves, their communities, and their institutions.'*   Howard Rheingold [372]

In the meantime, not only terrorism, but also cyber-related crime, has exploded to more than forty (40) percent of all criminal acts committed.[373] Due to the cyber-surveillance actions of the US government in 2001, over 150 organisations, 300 law professors, and 40 computer scientists have expressed support for the 'In Defense of Freedom statement'. Professor Kerry Kang of the UCLA School of Law said: 'Our constitutional tradition makes clear the need to proceed with the utmost regard for America's fundamental freedoms. We join this statement to express our commitment to fundamental civil rights, equal treatment under the law, and a deliberative process that guards against decisions we may later regret.'

New Rules about Data Mining

The right to privacy has in many societies been viewed as the cornerstone of freedom and liberty.[374] Freedom lies in the ability to better understand one's position in the world and to develop opinions independent of external pressures. Individual thought is considered a private matter. It has traditionally been thought that what one thinks, believes and knows is inalienable to oneself, and may only be revealed with the voluntary approval of the thinking person. In the past few decades, slowly and gradually, this notion has begun to erode. Today, all kinds of monitoring by all sorts of agencies seem to have become regular practice. Large amounts of information can be gathered by a variety of actors, for legitimate or illegitimate purposes. Some argue that the current concerns surrounding people's privacy result from a technological shift in communications, from one-way media to bi-directional flows of information. As such, it will become increasingly difficult to 'reveal without being revealed, and to learn without being learned about' (Stalder)[375].

---

[368] BBC News, 04-04-2001. WWW: <http://news.bbc.co.uk/1/hi/business/1260931.stm>.

[369] J Agar, Constant Touch: A Global History of the Mobile Phone, (Icon Books, Cambridge 2004).

[370] J van Kokswijk, 'The bionic buddy as remote control for life', Management & Informatie, 2001/6.

[371] L Srivastava, 'Mobile phones and the evolution of social behaviour', Behaviour & Information Technology, Vol.24/2 (2005).

[372] H Rheingold, *Smart Mobs, the next social revolution* (Perseus Publishing, 2002).

[373] Statement at the 5th International ICT-& Information Security Symposium; WWW: <http://www.govcertsymposium.nl/>.

[374] As the French essayist Michel de Montaigne said in the 16th century of the gap between the private and the public spheres of existence: 'A man must keep a little back shop where he can be himself without reserve. In solitude alone can he know true freedom', Michel de Montaigne, Essais, 1588.

[375] F Stalder, 'The Failure of Privacy Enhancing Technologies (PETs) and the Voiding of Privacy', *Sociological Research Online*, (2002) Vol. 7, no. 2.

In this context it has generally been accepted that data pertaining to the individual should only be propagated with the knowledge and consent of the individual concerned. Many governments and organisations (commercial or otherwise) show an awareness of this aspect by making disclaimers at the time of the acquisition of data. But these often voluntary efforts are feeble in the face of the many challenges present in this field. Bohn et al [376] state that because of these reasons, the privacy of personal data is actually of considerable importance, and moreover, its importance will grow with time. Nonetheless to some privacy prevention appears to be a trendy subject of only passing interest.

> **All about Alex** '*The privacy concerns ... refer to the protection of attributes and preferences associated with Alex's identity. He is only required to produce proof of age to purchase the alcohol. He is not required to disclose data such as the name of his college or the address of his employer. Moreover, as Alex paid in cash, neither his name, age nor license number were recorded. As such, Alex's early predilection for vodka will not be automatically communicated to his biology professor or to his parents. The privacy of his actions in this case is assured because the data in question is: a) minimal: only a driver's license was presented, b) temporary: the license was only examined briefly by the store clerk, and c) un-linkable: it cannot be linked with Alex's other attributes (parents' name and address or professor's contact details). These same considerations should apply to the online world, and indeed many proposed digital identity management schemes have focused on principles such as 'un-linkability' and data minimisation.*'   The Digital Life Report

Nowadays more and more governments around the world have become concerned about their ability to monitor the electronic lives of their citizens, residents and visitors. The amount of personal information generated in electronic form (phone calls, emails, SMS, data files, observation cameras, and so on) has grown exponentially, and this is making effective surveillance harder.

Nevertheless, using modern data mining tools, it is possible to combine information from many different sources to build up a detailed picture of the movements and habits of an individual person.[377] Software robots (bots) are increasingly used for collecting all kinds of individual information, not only by companies but also by governments. This can be a threat to privacy. [378]

Many of the hidden data collectors are unknown, but what is alarming is the already identified available information about massive surveillance activities of: [379]
- U.S. agencies, supported to some uncertain degree by U.K. agencies (the Echelon system, which targets at least European traffic, and captures and enables analysis of vast volumes of traffic);
- ENFOPOL (an emergent scheme involving police agencies in countries within the European Union); and
- A still-emergent French scheme (dubbed Frenchelon, in the absence of any known official title and which appears to be a Gallic reaction to the Anglo-Saxon Echelon, and which appears to capture far lower volumes).

This practise of profiling can be used to identify potential terrorists, but data mining had already intensively, and quietly, been used by market researchers who mostly mined the data from contacts and purchases in order to get a profile of the customer. In most situations the individual enabled this himself by accepting a loyalty smart card (for discounts) or by downloading a bonus feature that included spyware. In this way data mining became a powerful commercial tool for developing targeted marketing in a relatively short time. It is possible, for instance, to com-bine location-based information (e.g. from Internet access or mobile phone use) with information about the use of credit cards for purchases, to gain insight into when and where to send targeted messages to a potential customer. Each time we use an electronic device, pay or pass through a smart card, or give out our phone number or email address to a friend, we are surrendering a little bit of our own privacy.

If we have confidence in the way this information is captured, stored and used, and by whom, then we can decide to give up our privacy in return for added protection, or for more relevant services. But

---

[376] J Bohn, V Corama, M Langheinrich, F Mattern, and M Rohs, 'Living in a World of Smart Everyday Objects, Social, Economic and Ethical Implications', Human and Ecological Risk Assessment, Vol 10, No. 5, October 2004.

[377] R Clarke, 'Profiling: A Hidden Challenge to the Regulation of Data Surveillance'. *Journal of Law and Information Science* 4,2 (12/1993). WWW: <http//www.anu.edu.au/people/Roger.Clarke/DV/PaperProfiling.html>.

[378] BW Schermer, 'Software Agents, Surveillance, and the Right to Privacy: A legislative Framework for Software-Agent-Enabled Surveillance'. (PhD thesis Leiden University, 09-05-2007 Leiden University Press 2007); WWW: <https://openaccess.leidenuniv.nl/dspace/bitstream/1887/11951/1/Thesis.pdf>.

[379] Computers, Freedom, and Privacy 1999. 'The Global Internet. The Creation of a Global Surveillance Network' Panel Discussion proceedings. Held April 6, 1999. WWW: <http://cns.miis.edu/pubs/reports/terror_lit.pdf>.

once our trust in electronic networks is impaired, all aspects of our lives are at risk. In particular, our digital identity may be more vulnerable than we expect.

From Anonymity to Pseudonymity

The use of pseudonyms (also known as "nyms") enables the use of partial identities, and can thus cover the entire range from anonymity to identifiability. Pseudonyms allow users to take on different identities depending on the specific role, context and parties involved. The use of a "nym" is effective only when it cannot be linked to its holder or to other pseudonyms that a holder may have. Nonetheless, when necessary, the holder of the pseudonym can be revealed in many situations and as such, they could be liable and accountable for actions taken under that pseudonym.

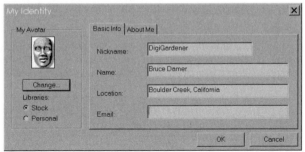

Figure 35: Log-in application screen with free choice of name, avatar et cetera.

May argues that when a government gives us a pseudonym (like a social security code) and forces us to use that, the government also must ensure that the name is a real name, especially because governments 'give themselves the rights to create/-forge completely false identities, such as for witness protection, undercover agents, etc.' [380]

It is proclaimed that digital identity management systems could handle both sides of the tendency to anonymity. Discussions regarding the principles which should be predicated are ongoing both nationally and internationally. Governments are becoming more concerned, particularly in an effort to thwart illicit actions and identity theft. The European Commission's approach to this question was first expounded in its PRIME (Privacy and Identity Management for Europe) project.[381]

In contrast Nabeth from The Centre for Advanced Learning Technologies (SALT) presented an overview of digital social environments from the viewpoint of the subject of identity.[382]
Its aim is to raise awareness on the diversity and richness of these environments, and on the different identity issues that may occur in these environments. Following Nabeth the conceptualisation of identity in DSEs (Digital Social Environments) relies to a large extent on a more informal and abstracted perspective of identity rather than the concept of identity manipulated by the information system or security specialists. Whereas in the latter case the identity is managed principally with people representations (as a set of attributes and characteristics that can be stored for instance in an identity card) and their authentication (usually taking place once at the beginning of a session), the management of identity in social environments is more diffused. In particular its control is not granted to a central authority but based on the idea of providing transparency about the behaviours and the actions of people, and is socially regulated (for instance with social pressure).

Nabeth argues that a closer look indicates that the two worlds (formal and informal perspective) are not totally alien but are on the contrary complementary, and are subject to cohabit more and more in the future, in particular as the identity issues in the Information Society are addressed more holistically. When the frontier between the physical and the digital worlds becomes blurred (this is best illustrated by the advent of ambient intelligent environments) and is converging, this new world becomes very complex and difficult to manage by traditional methods (via explicit identity and identifier, and one-time authentication), and as the Information Society becomes interested in more widely supporting human aspects (privacy, social dimension, etc.), new methods (in particular more flexible and more robust) will need to be activated in order to manage the identity in our societies.

---

[380] TC May, *True Nyms and Crypto Anarchy*, in: V Vinge and J Frankel (ed), 'True Names: And the Opening of the Cyberspace Frontier' (Tor Books 2001) pp 43 in pp 33-86.

[381] European Commission, PRIME White Paper. WWW: <https://www.prime-project.eu/prime_products/whitepaper/>.

[382] T Nabeth, 'Understanding the Identity Concept in the Context of Digital Social Environments' INSEAD CALT (the Centre for Advanced Learning Technologies) Paper was produced as part of the Network of Excellence FIDIS (Future of the Identity in the Information Society) a project of the 6th Framework programme of the European commission. WWW: <http://www.fidis.net/>.
FIDIS working paper, January 2005; WWW: <http://www.calt.insead.edu/Project/Fidis/ documents/ 2005-fidis-Understanding_the_Identity_Concept_in_the_Context_of_Digital_Social_Environments.pdf>.

Social mechanisms (reputation, social control, etc.) represent an effective means of regulation for complex systems, and should be considered (at least for a partial use) in every identity management solution. In particular, social engineering approaches to systems that include an important human component (including management of risk associated with inevitable human errors and biases), a category to which identity management systems belong to, are often more effective and more robust approaches than the technical engineering approach alone. Finally Nabeth indicated that the use of social mechanisms also has its limits, and that the management of identity in DSEs could benefit from the work of the more formal management of identity.

Jordan, Hauser, & Foster suggest that the social and formal approaches are complementary and should be combined in order to implement systems that are more flexible, more robust (in particular concerning human error) and more reliable, for designing the next generation of Internet systems that will be more socially aware, and more human friendly than they are today. [383]

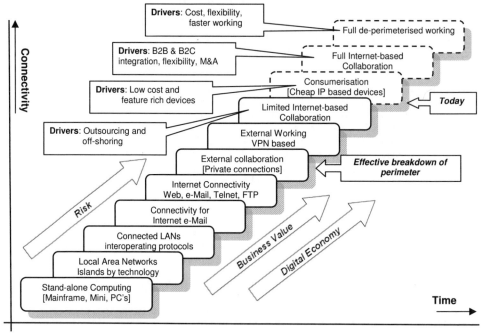

Figure 36: Computing history can be defined in terms in increasing connectivity over time, starting from no connectivity to the restricted connectivity we currently have today. Source: Jericho Forum.

Identification, Authentication and Data Surveillance

Looking at how people may be identified within technological systems, Roger Clarke argues that an identity is quite different from an entity. [384] A real-world entity, such as a human, may have multiple identities which represent aspects of themselves. In cyberspace as well as in the real world, there may be only one entity which has multiple identities associated with it, and as such, the issue becomes authentication rather than identification, checking that the presented data does indeed associate itself with the real-world entity that it says that it does. It goes further, suggesting that authentication need not reveal the entity behind the identity, but merely authorise access regardless of the entity behind the information. [385]

Most people believe that identity is simply one's name, age, gender and address. In fact, we all have multiple identities that are aspects of the entity which is the unique human being of flesh and blood

---

[383] K Jordan, J Hauser, and S Foster, 'The Augmented Social Network: Building identity and trust into the next-generation Internet'; *First Monday* 8/8; 04-08-2003. WWW: <http://www.firstmonday.org/issues/issue8_8/jordan/>.

[384] R Clarke, 'Human Identification in Information Systems: management challenges and public policy issues', *Information Technology & People*, 7(4) 1994) pp. 6-37.

[385] M Ito, 'Identity and Privacy in a Globalised Community', *Ars Electronica*, September 7-12 (Linz, Australia 2002) WWW: <http://www.aec.at/en/archiv_files/20021/E2002_244.pdf>.

that we are. (Burke)[386] Identities can be roles such as shareholder, officer, rape victim or spouse. Identities are identified by identifiers. Some identifiers require the authentication of the entity whereas some identities can be authenticated by uniforms, passwords, secret handshakes or other identifiers that do not expose the entity behind the identity. It is essential to consider the issue of identity independently from the issue of authentication of the entity.

The analogous situation in cyberspace is considered by Clarke through invoking the Jungian dualism of anima/persona, where the anima is the entity and there are digital personae created from their data trails. [387] He considers the creation of digital personae by individuals and the rise of data surveillance – particularly through matching personae on the basis of common identifier(s) – and profiling, where a set of characteristics is generated for particular purposes and data holdings are searched for digital personae that fit those characteristics. He identifies ten dangers to the individual and twelve dangers to society through the rise of data surveillance and its consequent chance of matching the digital persona to the real-life entity.

Laws of Identity

In response to these concerns about digital privacy and surveillance, 'Seven Laws of Identity' were formulated on an open blog (2003-2005) by a global community of experts under the leadership of Kim Cameron, Chief Identity Architect at Microsoft.

Cameron et al claim that the seven laws of identity 'define the set of 'objective' dynamics that constrain the definition of an identity system capable of being widely enough accepted that it can enable distributed computing on a universal scale'. [388] Cameron et al argued that people are sufficiently concerned about their digital identity to take steps to protect it and that only systems which adequately protect users and their identities will be widely adopted.

These laws are:

- **The law of control**. Technical identity systems must only reveal information identifying a user with the user's consent.

- **The law of minimal disclosure**. The solution which discloses the least identifying information is the most stable long-term solution.

- **The law of fewest parties**. Technical identity systems must be designed so that the disclosure of identifying information is limited to parties having a necessary and justifiable place in a given identity relationship.

- **The law of directed identity**. A universal identity system must support both omni-directional identifiers for use by public entities and unidirectional identifiers for use by private entities.

- **The law of pluralism**. A universal identity system must channel and enable the inter-working of multiple identity technologies run by multiple identity providers.

- **The law of human integration**. The universal identity system must define the human user to be a component of the distributed system, integrated through unambiguous human–machine communications mechanisms offering protection against identity attacks.

- **The law of contexts**. The unifying identity meta-system must facilitate negotiation between a relying party (a service that consumes the identity given in order to give access) and user of a specific identity presenting a harmonious human and technical interface while permitting the autonomy of identity in different contexts.

These seven laws, Cameron et al claim, form an objective set of criteria to which any system that seeks to be accepted as a backbone of identity-related distributed computing must conform in order to be widely accepted and adopted. The aim of the laws is to help users cut down on the degree to which data is shared and centralised.

> ... '*We have undertaken a project to develop a formal understanding of the dynamics causing digital identity systems to succeed or fail in various contexts, expressed as the Laws of Identity. Taken together, these laws define a unifying identity metasystem that can offer the Internet the identity layer it so obviously requires. They also provide a way for people new to the identity discussion to understand its central issues. This lets them actively join in, rather than everyone having to restart the whole discussion from scratch.*' ...   Kim Cameron [389]

---

[386] PJ Burke: See pages 23-24.

[387] R Clarke, 'The Digital Persona and its Application to Data Surveillance' *The Information Society*, 10(2) (1994).

[388] Kim Cameron's Identity Weblog. WWW: <http://www.identityblog.com/stories/2004/12/09/thelaws.html>.

[389] Channel9 Forums; Interview with Kim Cameron. June 2006. WWW: <http://download.microsoft.com/download/4/6/6/4667faf2-2915-422b-b91e-1373edd459a9/kim_cameron_on_identity_new_2005.wmv>.

In brief, the privacy-embedded Seven Laws of Identity, when implemented, offer individuals:
- easier and more direct user control over their personal information when online;
- enhanced user ability to minimise the amount of identifying data revealed online;
- enhanced user ability to minimise the linkage between different identities and actions;
- enhanced user ability to detect fraudulent messages and Websites, thereby minimising the incidence of phishing and pharming.

The Information and Privacy Commissioner of Ontario (Cavoukian) recently published a plan for automated internet privacy that recognises and is inspired by the Cameron's laws. [390]
Just as the Internet emerged from connecting different proprietary networks, an 'Identity Big Bang' is expected to happen once an open, non-proprietary and universal method to connect identity systems and ensure user privacy is developed, in accordance with universal privacy principles. Therefore Cavoukian has called for programmers to embed privacy capabilities in software, which implicitly means that virtual identities will acquire a main role in the software development architecture.

De-perimeterisation

De-perimeterisation is a concept/strategy used to describe protecting an organisation's systems and data on multiple levels by using a mixture of encryption, inherently-secure computer protocols, inherently-secure computer systems and data-level authentication rather than the reliance of an organisation on its (network) boundary to the Internet. De term is coined by Jon Measham in a 2001 research paper. [391]
Traditional security solutions like network boundary technology will continue to have their roles, but we must respond to their limitations. In a fully de-perimeterised network, every component will be independently secure, requiring systems and data protection on multiple levels, using a mixture of encryption, inherently-secure computer protocols, inherently-secure computer systems, and data-level authentication. Successful implementation of a de-perimeterised strategy within an organisation thus implies that the perimeter, or outer security boundary, could be removed. The design principles that guide the development of such technology solutions capture the essential requirements for IT security in a de-perimeterised world. This is defined by the Jericho Forum in their 'Commandments'. [392]

Summary Chapter 6: *'Virtual Identity and Anonymity'*

Underestimating the powerful features of the World Wide Web and staggered about the rise of virtual citizens, governments have failed to work out concrete policies for the Internet. The USA rushed rules for surveillance, the EU halted at a roadmap to build and strengthen the EU research community in this area, and the UN identified a vacuum within the context of existing structures, since there is no global multi-stakeholder forum to address Internet-related public policy issues.

The Internet and telephony lead to the establishment of virtual contacts on a large scale on the basis of interests or involvement, whereby each contact in itself is part of one of the many groups (so-called communities of interest), where practise shows that they are using different kinds of identities, with or without an URI or personal IP address. Multiple proposals have been made to build an identity management infrastructure into the Web protocols. All of them require an effective public key infrastructure so that the identities of separate manifestations of an online identity are probably one and the same. The Laws of Identity and the Jericho Forum are guiding us. The future of online anonymity depends on how an identity management infrastructure is developed.
In the next chapter the legal position of the virtual identity is worked out.

---

[390] A Cavoukian, '7 Laws of Identity: The Case for Privacy-Embedded Laws of Identity in the Digital Age.' (Ontario, 2006). WWW: <http://www.ipc.on.ca/images/Resources/up-7laws_whitepaper.pdf>.

[391] Valueless Security. WWW: <http://www.opengroup.org/projects/jericho/uploads/40/5368/Value-Less_Security_v3.pdf>.

[392] The Jericho Forum is an international IT security thought-leadership group dedicated to defining ways to deliver effective IT security solutions that will match the increasing business demands for secure-IT operations in our open, Internet-driven, globally networked world. Multi-national corporate user organizations, major security vendors, solutions providers, and academics are working together to concrete a de-perimeterised future. WWW: <http://www.opengroup.org/jericho/>. The commandments define both the areas and the principles that must be observed. WWW: http://www.opengroup.org/jericho/commandments_v1.1.pdf>.

7 Legal Issues and Policy of Virtual Identity Part I

'It cannot be helped; it is as it should be, that the law is behind the times.'

Oliver Wendell Holmes (1841-1935).[393]

Introduction
Since the beginning of humanity, certain values and norms have existed, and almost the same long-standing discussion on the regulation is adequate. Ancient also is the need of people to leave their regular identity in order to embody themselves in imaginary worlds. As technology shapes this virtuality and society moves to a virtual world the need exists to regulate consequential human behaviour such as transactions, awful behaviour and unwanted acts. The virtuality mostly arose in the globally connected computer networks, usually called the Internet, with the idea positioned that 'virtual' means supernal and intangible. Popular are the discussions concerning jurisdiction and legality, as well as the use of all kind of rules in order to regulate human behaviour with the purpose of making an individual fit into society. On the basis of the global evolution of norms, order, rules, and the law, we will consider in this chapter the policy around the legal issues of virtual identity.

The Crawl of Codes
This subchapter points out the history of various systems of law, both national and exterritorial, and focuses on the different theories about sovereignty and jurisdiction.

The Progress of Order and Law

Thou Shalt Not Make for Yourself an Image
Aspiratory statements are likely to provide the best evidence of rules in ancient times. For example, the Ten Commandments, and equivalent aspiratory statements in the sacred texts of other cultures, provide evidence of rules that prevailed in antiquity. The Ten Commandments, or Decalogue, are a list of religious and moral imperatives which, according to Biblical tradition, were written by God and given to Moses on Mount Sinai in the form of two stone tablets. They feature prominently in Judaism and Christianity.

> Commandment 3: '*You shall not make for yourself an image, whether in the form of anything that is in heaven above, or that is on the earth beneath, or that is in the water under the earth.*' (1300 BC) [394]

Some years later...

> '*As a god, I can be a devil. When I got up on the wrong side of bed on a recent Sunday, I took it out on my little worshipers in the computer game Black & White. I sought to create rock slides on mountainsides above their island community. When that proved ineffective, I dropped boulders and trees directly on their houses. When the roofs crashed in and the occupants ran screaming from the wreckage, I picked them up and threw them about like dolls. Then I dropped rocks on them, too. (...) As it turns out, I can be a moody, unpredictable god, and when I have a tantrum, I am not shy about who knows it.*' (2001 AD) [395]

---

[393] Oliver Wendell Holmes, *Speeches 1934*,(Brown & Co. Little, Boston 1934) p. 102.

[394] The first edition of the Ten Commandments is referenced in the History of the Law page, as part of Lloyd Duhaime's LAW Museum Archives collection. The phrase 'Ten Commandments' generally refers to the broadly identical passages in Exodus 20:2-17 and Deuteronomy 5:6-21, but is used by some to refer to the so-called 'Ritual Decalogue' (Exodus 34). According to the Bible, it was in approximately 1300 BC that Moses received a list of ten laws directly from God. These laws were known as the Ten Commandments and were transcribed as part of the Book of Moses, which later became part of the Bible. Many of the Ten Commandments continue in the form of modern laws such as 'thou shalt not kill' (modern society severely punishes the crime of murder), 'thou shalt not commit adultery' (modern society allows a divorce on this grounds) and 'thou shalt not steal' (modern society punishes theft as a crime). In some versions 'image' is replaced by 'idol'. The Bible chapter that contains the Ten Commandments (Exodus) follows the recitation of the Commandments with a complete set of legal rules, which are based on the 'eye for an eye, tooth for a tooth' legal philosophy of Hammurabi's Code; WWW: <http://www.duhaime.org/Law_museum/tencomm.aspx>.

[395] P Olafson, 'Review of the 2000 Computer God Game "Black & White": A deity in Touch With His Own Bad Self' *New York Times*, April 5, 2001. There have been other god games: Afterlife, Pharaoh (Build a Kingdom, Rule the Nile, Live Forever); Zeus: Master of Olympus, etc. WWW: <http://freenet-homepage.de/maude/God-Games.htm>.

'The Laws' [396] is Plato's last and longest dialogue. The question asked at the beginning is not 'What is law?' as one would expect – that is the question of the Minos. The kick-off question is rather, 'Who is given the credit for laying down your laws?'

It is generally agreed that Plato (427? BC - 347? BC) wrote this dialogue as an old man, after having failed in his effort to guide a tyrant's rule in Syracuse on the island of Sicily, and having himself thrown into prison instead.

The questions of 'The Laws' are imagined [397] without limit:
- Divine revelation, divine law and lawgiving,
- The role of intelligence in lawgiving,
- The relationships between of philosophy, religion, and politics,
- The role of music, exercise and dance in education,
- Natural law and natural right.

Plato described in 'The Laws' the governance of people by the word 'govern' [398]. This 'govern' concerns the theory of teleology: form follows function. Example: a person has eyes because they have the need of eyesight. Opposite is the theory of naturalism that would say that a person has sight simply because they have eyes. Function follows form. In other words: eyesight follows from having eyes.
Elaborating on this theme Plato sets out in 'The Republic' [399] the question to define: 'What is justice?'. Given the difficulty of this task, Socrates and his interlocutors are led into a discussion concerning justice in the city, which Socrates suggestions may help them to see justice in the person, but on a grander (and therefore easier to discuss) scale.

In the following centuries– retroactively, incident by incident, subsequently, and sometimes simultaneously – all kinds of code, from 'the law of the forefathers' to today's international private law, were created all over the world, both in and ex territories. Within the scope of this survey it goes too far and too deep to expose the development of the law. For those studying the past and/or the law, Benson's 'Where Does Law Come From?' could be interesting to read. [400]

Significant is the fact that there is also international 'canon law'[401], especially related to imagining. [402] The Catholic Church has a long history about imagined societies and superterrestrial persons. [403] US priest and Professor Andrew Greeley identifies the Catholic imagination as sacramental; an imagination that recognises the incarnation of God's presence in creation: [404]

> 'Catholics live in an enchanted world, a world of statues and holy water, stained glass and votive candles, saints and religious medals, rosary beads and holy pictures. (...) As Catholics, we find our houses and our world haunted by a sense that the objects, events, and persons of daily life are revelations of grace.'

---

[396] Plato, The Laws. (360 BC) Translation: B. Jowett WWW: <http://classics.mit.edu/Plato/laws.html> and: Plato, Dialogues, Laws, Index to the Writings of Plato.V5 WWW: <http//oll.libertyfund.org/Home3/Book.php?recordID=0131.05>.

[397] See especially: MW Bundy, 'Plato's View of the Imagination' Studies in Philology 19 (1922) pp 362-403; also: 'The Theory of Imagination in Classical and Medieval Thought', Studies in Language and Literature, University of Illinois 12 (1927) pp 2-3; also G Watson, Phantasia in Classical Thought (Galway University Press, Galway 1988).

[398] Govern was used in cybernetic theories for the governed engines, to achieve a kind of interaction between the human and the machine, and since virtual reality also as enabler for more realistic imagination. The study of teleological mechanisms (from the Greek τέλος or telos for end, goal, or purpose) in machines with corrective feedback dates from as far back as the late 1700s when James Watt's steam engine was equipped with a governor, a centrifugal feedback valve for controlling the speed of an engine. In 1868 Maxwell published a theoretical article on governors. In 1935 Russian physiologist Anokhin published a book in which the concept of feedback ('back afferentation') was studied. The Romanian scientist Odobleja published Psychologie consonantiste (Paris, 1938), describing many cybernetic principles. In the 1940s the study and mathematical modelling of regulatory processes became a continuing research effort and two key articles were published in 1943. These papers were 'Behaviour, Purpose and Teleology' by Rosenblueth, Wiener, and Bigelow; and the paper 'A Logical Calculus of the Ideas Immanent in Nervous Activity' by McCulloch and Pitts. In 1948 Wiener presents his 'Cybernetics: Control and Communication in the Animal and the Machine'.

[399] Plato, The Republic (360 BC) Translation: B. Jowett; WWW: <http://classics.mit.edu/Plato/republic.html>.

[400] BL Benson, 'Where Does Law Come From?' In: The Freeman, December, 1 1997; WWW: <http://www.independent.org/newsroom/article.asp?id=202>.

[401] Canon Law, the juridical law of the Roman Catholic Church, is not a moral code, it is the administrative, civil, jurisdictional, procedural and penal law of the Catholic Church. It is subject to many of the same political, administrative and practical influences that shape any other body of law; while the authoritative moral teachings of the Church belong to a different forum.

[402] J Coriden, R Pagé, and R Torfs, 'Canon Law Between Interpretation and Imagination' W Onclin Chair 7/2001.

[403] BJ Kelty, 'A Vision for Catholic Education in the Twenty-first Century' (2001); WWW: <http://dlibrary.acu.edu.au/research/theology/ejournal/Issue2/BRIAN_Kelty.htm>.

[404] A Greeley, The Catholic Imagination (University of California Press, Berkeley CA. 2000) p 1.

A catholic education feeds this imagination by its development of the imagination especially through exposure to literature, art, history and the classics and canon law regulates imagination. Catholics are not considered to imagine something that is out of their consideration.
As Greeley argues e.g.: '... whether Jesus felt erotic attraction toward women, as he is not only God but also human, one is warned that the question is improper'. Nonetheless, he concludes that 'Catholics are less likely than Protestants to support obedience to laws; Catholics are more likely than Protestants to approve violent protests', however a study shows that they realize that their salvation depends on whether they adhere to the (canon) law perfectly. [405]

Other religions also have a kind of canon law, like the Islam (*Shariah* [406]).

Canon law, on the whole, is more exceptional as it is a global law, so a-territorial. And – in specific parts – it concerns unreal personalities. That is what makes it interesting in the perspective of virtual identity.

In many aspects the cyberworld is characterised by ancient Roman ways of acquiring sovereignty such as the way of acquisition, law making and lack of real enforcement.
Austin [407] states that 'every successive Emperor, or every successive Prince, could acquire the virtual sovereignty of the Roman Empire or World' without having any testament, nomination, appointment, election or what else. 'Every successive Emperor acquired by a mode of acquisition which was purely anomalous or accidental.' ([407]: p 152). He argues that 'Laws set by *general* opinion, or opinions or sentiments of *indeterminate bodies*, are only opinions or sentiments that have gotten the name of *laws*. But opinions or sentiment held or felt by an *individual*, or *all* members of a *certain aggregate*, may be as closely analogous to a law.' ([407]: p 154). However: '...The term *law*, or *law set by opinion*, is never or rarely applied to a like opinion or sentiment of a precisely determined party: that is to say, a like opinion or sentiment held of felt by an individual, or held or felt universally by all the members of a certain aggregate.' ([407]: p 155). This situation seems equal to some cyber regulating.

Qua Lege Vivis? According to what law are you living?
Originally, law had a non-territorial basis. (Johnsson) [408] The principle of territorial sovereignty was unknown in the ancient world. In fact, during a large part of what we usually term modern history, no such conception was ever entertained. In the earlier stages of human development, ethnicity or nationality rather than territory formed the basis of a community of law. During this period, an identity of religious worship seems to have been a necessary condition of a common system of legal rights and obligations. The barbarian was outside the pale of religion, and therefore incapable of amenability to the same jurisdiction to which the natives were subjected (Johnsson). For this reason, in the ancient world foreigners were either placed under a special jurisdiction or completely exempted from the local jurisdiction. In these arrangements for the safeguarding of foreign interests we find the earliest traces of extraterritoriality. Extraterritoriality originally was a system of non-territorial governance. The laws followed the person, instead of the territory. Thus, in one and the same place, people could submit to various systems of laws.

---

[405] D Dudley and VB Gillespie, Value genesis data, Section on grace and works (La Sierra University Press, Riverside, CA 1992) WWW: <http://www.lasierra.edu/centers/hcym/media/images/pdfs/vg2-update-v3.pdf>.

[406] Sharia, the religious law of Islam. As Islam makes no distinction between religion and life, Islamic law covers not only ritual but every aspect of life. The actual codification of canonic law is the result of the concurrent evolution of jurisprudence proper and the so-called science of the roots of jurisprudence (usul al-fiqh). A general agreement was reached, in the course of the formalization of Islam, as to the authority of four such roots: the Qur'an in its legislative segments; the example of the Prophet as related in the hadith; the consensus of the Muslims (ijma), premised on a saying by Muhammad stipulating 'My nation cannot agree on an error'; and reasoning by analogy (qiyas). Another important principle is ijtihad, the extension of sharia to situations neither covered by precedent nor explicable by analogy to other laws. These roots provide the means for the establishment of prescriptive codes of action and for the evaluation of individual and social behavior. WWW: <http://www.answers.com/topic/sharia>.

[407] J Austin, *The Province of Jurisprudence Determined and the Uses of the Study of Jurisprudence* (Hackett Publishing Company, 1832/1863; Repr edn 1998).

Austin differentiates: 'Laws proper, or properly so called, are commands; laws which are not commands, are laws improper or improperly so called. Laws properly so called, with laws improperly so called, may be aptly divided into the four following kinds:

1. The divine laws, or the laws of God; that is to say, the laws which are set by God to his human creatures.

2. Positive laws, that is to say, laws which are simply and strictly so called, and which form the appropriate matter of general and particular jurisprudence (…) may be styled laws or rules set or imposed by opinion.

3. Positive morality, rules of positive morality, or positive moral rules.

4. Laws metaphorical or figurative, or merely metaphorical or figurative.

[408] Non-territoriality has also been called exterritoriality, a-territoriality or extraterritoriality. See: RCB Johnsson, 'Non-Territorial Governance – Mankind's Forgotten Legacy' (2005); WWW: <http://www.panarchy.org/johnsson/review.2005.html>.

> 'In the absence of any views of territorial sovereignty, there developed in medieval Europe a complete system of personal jurisdiction, which has left in its wake many interesting survivals extending to modern times, and which undoubtedly exercised an immense influence upon the development of extraterritoriality. In the days which followed the downfall of the Roman Empire, as in the days of ancient Greece and Rome, but in a much more marked degree, racial consanguinity was treated as the sole basis of amenability to law. Thus, in the same country – and even in the same city at times – the Lombards lived under Lombard law, and the Romans under Roman law. This differentiation of laws extended even to the various branches of the Germanic invaders; the Goths, the Franks, the Burgundians, each submitted to their own laws while resident in the same country. Indeed, the system was so general that in one of the tracts of the Bishop Agobard, it is said: 'It often happens that five men, each under a different law, would be found walking or sitting together.' (p. 27).
>
> 'During the sixteenth and seventeenth centuries, an era of dynastic and colonial rivalry set in. The discovery of America initiated among the more powerful maritime Powers of Europe the struggle for colonial possessions. The ascendancy of these Powers aided their assertion of an exclusive territorial sovereignty, until 1648 the treaties making up the Peace of Westphalia accepted the latter as a fundamental principle of international intercourse. This development of territorial sovereignty was distinctly fatal to the existence of the system of consular jurisdiction [i.e. the extraterritorial courts], and facilitated considerably its decadence in Europe, because it was founded on the opposite theory of the personality of laws' (p. 37). Liu [409]

It seems that the Peace of Westphalia in 1648 [410] was a decisive year, as the idea of exclusive territorial sovereignty replaced the theory of the personality of laws as the fundamental principle of international intercourse. After the development of modern cartography, legal authority generally followed relation-ships of status rather than those of autochthony. We are now accustomed to territorial jurisdiction so much so that is it hard to imagine that in past the jurisdiction was a-territorial. 'It is a recognised principle of modern international law that every independent and sovereign State possesses absolute and exclusive jurisdiction over all persons and things within its own territorial limits' (Liu, pp 17, 27, 37)

Comparing with the situation of the virtual worlds, where – from a user perspective – state boundaries are less important than in the physical world, there could be thoughts about customary law with private means of resolving disputes and dispensing justice.
Benson (2005)[411] describes a modern system of law and order without state coercion that could be useful. Despite the desirability, countries simply are not willing to give up their sovereignty simply because their citizens participate in global cyberspace. Years of negotiation have not even solved jurisdictional issues, so it is not surprising that they cannot produce harmonisation of cyber law. (Newman)[412]
Even if jurisdictional issues could be resolved and agreements could be reached by governments regarding the appropriate rules to enforce, individuals who want to break such laws in cyberspace would be able to dramatically limit the likelihood of detection and punishment. After all, they can hide in cyberspace for a long time.

Sovereignty and Jurisdiction

Borderless Law and Jurisdiction

Virtual identities are human harebrained schemes. Similarly, so is the entity of 'legal persons'. Favourable is that man can create almost everything. Inconvenient is that many creations result in artificial fabrications that are so vague that the users of the creature are amused but the non-users want to regulate the specific phenomenon. That divide can exist without serious trouble until the moment that one party obstructs another party. In a standoffish environment people will battle out their dispute but in a social circle of acquaintances people will try to resolve the quarrel. Social control can be a really good guide in settling the disputes and achieving commitment. Ellickson

---

[409] S Liu, *Review of Extraterritoriality: Its Rise and Its Decline* (Columbia University Press, New York 1925) p 17-37. WWW: <http://www.panarchy.org/shihshunliu/presentation.1925.html>

[410] Treaty of Westphalia; 24-10-1648. WWW: <http://www.yale.edu/lawweb/avalon/westphal.htm>.

[411] B Benson, 'Customary Law with Private Means of Resolving Disputes and Dispensing Justice: A Description of a Modern System of Law and Order without State Coercion', *The Journal of Libertarian Studies*, Vol. IX, No. 2 (Fall 1990) WWW: <http://www.mises.org/journals/jls/9_2/9_2_2.pdf>.

[412] M Newman, 'E-Commerce (A Special Report): The Rules --- So Many Countries, So Many Laws: The Internet May Not Have Borders; But the System Certainly Does,' p 4. *Wall Street Journal*, April 28, 2003; Eastern edition, Oct 20, 2003; R.5.

shows examples of people governing themselves largely by means of informal social rules without the aid of a state or other central coordinator in his work 'Order without law' [413].

In the popular perception, a "gold rush" refers to a chaotic scramble for high-profit opportunities in an open-access setting, where the premium is on speed. Among law-and-economics specialists, the mining districts of the California gold rush (>1848) are often cited as canonical examples of spontaneous establishment of secure property rights in the absence of legal authority. There are more similarities between the gold colonists and the cyber-squatters. Both developed in their colonial period their own tailor made code, used primitive tools in the beginning, were dependent on imported propriety technologies and caused enormous changes to the region's culture and architecture.

Gold rush order

'When the Spanish arrived in northern California in the 1770s, they brought building traditions and technologies from Europe and Mexico. Colonial structures were larger and more complex than those of Native Californians. They required both the craftsmanship of skilled artisans and physical labor provided by Spanish and Native workers, and animals. Like Native Californians, colonial settlers relied primarily on local materials. The most prevalent type of construction used adobe bricks, made from clay and straw, and a masonry technology common in Spain and Mexico. Walls were constructed of adobes held together with mud mortar. Buildings had flat or pitched roofs, made of clay tile or thatching, and were often covered with a white limestone coating. More substantial buildings were constructed of stone from local quarries.

The economic development of mining, hydraulic engineering, hydroelectric power, and transportation during the late-nineteenth and early-twentieth centuries trans-formed the California landscape. These technologies provided strong ties between urban and rural areas, forging a permanent link between industrial development and population growth in cities, and the natural resources that supported them. The physical features of this infrastructure -- bridges, power lines, dams, aqueducts, and roads -- have become conspicuous features of the modern landscape.

No federal mining law was in existence at the time gold was discovered. 'By another coincidence, Marshall's discovery came just two years after an important turning point in national minerals policy. Frustrated by widespread non-compliance and fraud in its attempts to gain revenue from lead and copper mines, the federal government abandoned all administrative apparatus and enforcement machinery pertaining to minerals on the public domain in 1846. For some months, gold mining went forward under truly wide-open conditions, subject to no regulation of any kind. This state of affairs could not last, however. Increased population in the mines, particularly after mid-1849, created demand for some type of order.' [414]

The first change was the emergence of the idea of a claim. Legal historian Andrea McDowell shows that the concept of a 'claim' as an area of land as opposed to a hole in the ground did not become standard until 1849, though there were scattered uses of the term earlier. Within a matter of months, however, some basic rules became widely accepted, which McDowell calls the 'common law or customary law of the diggings'[415]. McDowell quotes from an account by miner Felix Paul Wierzbicki, written in 1849: 'A tool left in the hole in which a miner is working is a sign that it is not abandoned yet, and that nobody has a right to intrude there, and this regulation, which is adopted by silent consent of all, is generally complied with.'

Perhaps the most fundamental of these rules was that tools left in a hole indicated that the miner was still actively mining, and so the hole and the immediately adjacent land should not be interfered with. Soon after the idea of a claim, we see miners meeting in order to establish the law of the land for a geographical area, the mining district. Umbeck refers to the mining codes as 'contracts'. [416] When we "teleport" this idea to the virtual world, the analogy can be that the virtual tools left in a part of cyberspace prove the activity and property of that virtual space. In case of theft, the logging of the particular server could be used as proof of the presence (and action) of each "trespasser". MacDowell argues that district rules were not contracts in any standard sense, but agreements among a list of signers to respect and enforce each other's rights. Mining district codes were 'laws of

---

[413] RC Ellickson, *Order without law. How Neighbors Settle Disputes* (Harvard University Press 1991).

[414] K Clay and G Wright, 'Order Without Law? Property Rights During the California Gold Rush' (2003) WWW: <http://www-econ.stanford.edu/faculty/workp/swp03008.pdf>.

[415] AG McDowell, 'From Commons to Claims; Property Rights in the California Gold Rush' *Yale Journal of Law and Humanities* (2002) pp 14-15.

[416] J Umbeck, 'The California Gold Rush: A Study of Emerging Property Rights', *Explorations in Economic History* 14 (1977) pp 197-226.

the land' for a specified area, rules and procedures binding to all miners in that district, founding members and newcomers alike.

Lawyer Henry Halleck wrote in 1860: 'The miners of California have generally adopted as being best suited to their particular wants, the main principles of the mining laws of Spain and Mexico, by which the right of property in mines is made to depend upon *discovery* and *development*; that is, *discovery* is made the source of title, and *development*, or working, the condition of continuance of that title. These two principles constitute the basis of all our local laws and regulations respecting mining rights.' [417] Those principles can rule for the exploring of the cyberworlds too.

In the past, various commentators have suggested analogies for the analysis of law in cyberspace, including the 'wild west' and the feudalism. However, the best analogy is international commerce, since this operates under a polycentric system of customary law. (Johnson and Post, p: 1389-1390) [418] Benson [419] argues that evolving property rights, contracting arrangements (sources of trust and recourse), and customary law in cyberspace can provide effective governance in cyberspace.

Cross-border Law and Jurisdiction

In his year of death 1268 the English jurist De Bracton left behind him a manuscript that contained just one passage on jurisdiction, which was that 'Jurisdiction is nothing else than having the authority to declare the law or to adjudicate between parties in actions, touching persons or themes according as they are brought into court.'[420] De Bracton's simple description of adjudicatory jurisdiction remains valid today. It is the central concept for ordering the application of laws in space. It is concerned with the application of laws in space and how one determines which space the law shall apply to.

Since the 13th century it is not De Bracton's definition which has changed but the organisation of the space to which the law has to apply. In the medieval times there were competing potentates within the jurisdiction of England. Jurisdiction was a concept which was used to balance the civil jurisdiction, the ecclesiastical jurisdiction and the various powers and, indeed, jurisdiction was the princepal means by which the law was used as the means of establishing the authority of the sovereign. This eventually becomes what we now know as the common law for common law countries.

Today we live in a system in which the international legal order is based on separate physical jurisdictions, each with the sovereign attributes of prescriptive jurisdiction, the ability to make the law, the adjudicative jurisdiction, the ability to declare the application of the law, and the power of enforcement.

However, we are increasingly being confronted with the economic aspects of global markets and with the technological means of global communication, such as the Internet. We also are confronted with a very radical disjunction between economic and technological realities which are global and legal realities as well as national. The major question in this world of disjunction is: Is the significance of jurisdiction in that context merely a technical mechanism for determining the application of law, or is it the expression of political power?

In the following **Cases** it will be become clear only laws and political power are competing.

Arbejdsret

Law cannot wait for diplomacy as jurisprudence takes care of progress in lawmaking. A nation's power to enforce its laws is limited by territory. As long as the act and the loss happen in the same country, the legal case can be relative simple. Almost every country has a penal code about tort in some kind of civil law, and the court will judge the wrongful act and pass sentence on compensation from offender to victim. When more countries are involved the jurisprudence shows situations of 'borderless law'. If we compare events on a server with events on board of a ship, the jurisprudence indicates that the contracting state to which the ship (server) is connected must necessarily be

---

[417] HW Halleck, 'Introductory Remarks by the Translator' in JHN de Fooz, Fundamental Principles on the Law of Mines. (San Francisco, CA. 1860).

[418] DR Johnson and D Post, 'Law and Borders - The Rise of Law in Cyberspace,' Stanford Law Review 48(1996) pp 1367-1402.

[419] BLB Benson, 'Jurisdictional Choice in International Trade: Implications for Lex Cybernetoria, *Journal des Economistes et des Etudes Humaines*, 10 (2000) pp 3-31.

[420] H de Bracton, *De legibus et consuetudinis Angliae*, in: GE Woodbine (ed) 4 vol. (New York 1915-1942); re-edition with translation and revisions by SE Thorne, 4 vol. (Cambridge, Mass. 1968-1977; reprint 1997); WWW: <http://bracton.law.cornell.edu/bracton/Common/index.html>.

regarded as the place where the harmful event caused damage if the damage concerned arose aboard the server in question. (Arbejdsret case) [421]

Potash Mines

In the potash mines water pollution decree the EU-court clarified the term 'location where the fact of damages occurs': [422]

> In the potash mines case a food nursery was suffering a loss by the salt pollution of the water in the river Rhine. Research indicates that at least the river was polluted by the waste water from the potash mines in France, approx. 500 km's upstream. Also there were indications from draining of chemical industries along the river in Germany and Switzerland. The pollution was taken along downstream and when meeting the crosscurrent of seawater in the delta the pollution sinks to the bottom of the river in the environment of Rotterdam. This not only implicates bad food and sick cows when drinking the surface water from the river but also high concentrated sediments at the bottom of the river and problems with the drinking-water supply. The city of Rotterdam still has huge costs to provide healthy potable water for their citizens. [423]

Circuit Case

One of the most remarkable examples of jurisprudence is the judgment in cases where more parties (people; companies; scenes) are involved. Each party will have a share in the final result (such as a value chain) and thus they could be connected to/by a 'circuit'. In such situations each (legal and/or natural) person in the circuit is expected to know that this person was part of/in that circuit, was alleged to belong to a suspected group, or at least has tolerated (the connection to) that circuit. Each party of such a circuit can be punished by criminal justice and by a civil claim. For more criminal situations (such as criminal syndicalism) there are definitions of the concept of organised crime.[424]

Thalidomide

Also, one party can be condemned to pay the penalty and/or the given damages, even when there could be more suspects that have not been captured and judged.
A well known circuit related case in jurisprudence is the thalidomide case. [425] In a suit against Richardson-Merrell, the drug company that tested thalidomide in the United States in 1960 and

---

[421] ECR 1735 Danmarks Rederiforening acting on behalf of DFDS Torline A/S v LO Landsorganisation i Sverige, acting on behalf of SEKO Sjofolk Facket for Service och Kommunikation. Case C-18/02 (25 January 2002); WWW: <http://eur-lex.europa.eu/LexUriServ/LexUriServ.do?uri=CELEX:62002J0018:EN:HTML>.

[422] Handelskwekerij GJ Bier B.V. and others v Mines de Potasse d'Alsace S. ECR [1976] 1735. 4. 148/78 (30 November 1976) Case 21-76; WWW: <http://eur-lex.europa.eu/LexUriServ/LexUriServ.do?uri=CELEX:61976J0021:HU:NOT>.

[423] NN, 'De prijs van het water', Dutch analysis about the costs of water management in the Netherlands (2002);WWW: <http://www.lei.dlo.nl/publicaties/PDF/2002/3_xxx/3_02_01.pdf>.

[424] The expansive definitions of the EU and the UN reflect a political dimension of the concept of organised crime. It is a new "buzzword" in the policy-making community and it is perceived as a 'prime threat' to societal security alongside terrorism after 11 September 2001. The EU definition consists of four obligatory characteristics and two characteristics from a list of seven optional features that a crime must satisfy in order to be considered 'organised crime'. The obligatory features are:

a. the presence of more than two perpetrators;

b. criminal activity stretching over a long period, namely the stability of the criminal operation or group activity;

c. the commission of 'serious offences', i.e. any offence that is punishable by an upper limit of four years of imprisonment or above, according to the relevant national criminal law;

d. the main motive of the criminal activity is illicit profit or the attainment of (political) power.

The list of optional features includes the following:

a. a specialist division of labour and distribution of tasks among the participants in the criminal enterprise;

b. discernible mechanisms of discipline and control within the criminal group;

c. the use or threat of violence in the conduct of criminal activity;

d. the use of commercial or business-like structures;

e. money-laundering activity;

f. cross-border or international criminal activity;

g. the exertion of influence over legitimate state institutions.

The satisfaction of any two of the optional criteria plus all four obligatory criteria renders a criminal offence "organised crime" and qualifies it for special treatment as provided for by national and international legislation and policy.

The UN definition is similar and is given in Article 2 of UN Convention Against Transnational Organised Crime.

[425] Case Richardson-Merrell Inc. v Koller et al. Supreme Court of the US: 26-02-1985, No. 84-127; WWW: <http//www.oyez.org/cases/case/?case=1980-1989/1984/1984_84_127>.

1961, the judgment was against one company, even though more had manufactured the same drug. The company should find themselves their 'mates in businesses' in order to share equally the penalty and the given damages.

> Thalidomide, which was marketed as a sleeping aid (with names such as Softenon), was banned worldwide after 12,000 babies were born with serious birth defects, including flipper-like arms, to women who had taken the drug. The company's lawyers argued that it could not be held responsible because the drug had been taken too late in the pregnancy to cause harm. But it was countered that experts might be mistaken about what stage of pregnancy was critical to a baby's development. The jury returned a $2.75 million verdict, later reduced to a reported $500,000.

Scientology

In the Scientology case the preliminary decision of the Amsterdam District Court of 9 June 1999 included a separate declaratory judgment stating that providers must take action if they are made aware of material on their servers that infringes upon a copyright if 'the correctness of the notification of this fact cannot be reasonably doubted'. This decision also made reference to hyperlinks to material that infringes upon a copyright. According to the decision, if a provider is aware of this, it must take action against these hyperlinks. From the business driven Internet service provider 'certain carefulness can be expected regarding preventing further violation'. [426] This was largely confirmed by the Dutch Court of Appeal.[427]

DB Indymedia

In the '*Deutsche Bahn vs. Indymedia*' [428] judgment about linking from one Website (hosted in The Netherlands by provider XS4ALL) to another foreign Website that published an article with various possibilities to sabotage the German railway company, the Court defined that a link to unlawful information also is unlawful.

> Deutsche Bahn, the German rail company, won a court case against Indymedia.nl, ordering Indymedia.nl to remove all direct and indirect links to two pages that contain information considered unlawful in the Netherlands. Indymedia.nl argued that they never linked directly to the pages, such as inline or frame linking. It linked to several copies of the home page of an archive, some of them containing the incriminated articles. However, not all of these sites even fell under the Dutch jurisdiction. Also, removing all indirect links would literally mean removing all links whatsoever on Indymedia.nl, since every page on the Internet can be reached by any other page.

> Referring to the Scientology case in 1996, the Provisional Court argues: (translated JvK) ` The question which form of hyper-linking is used becomes irrelevant in this context. Decisive is that Indymedia makes it technically possible and kept the information at reach. Whether that occurs indirectly or directly is not important. In this context this rules to remove now the accompanying texts of the hyperlinks as they also explicitly call on the reader to unlawful behaviour and thereby give them the required instructions to execute it. The circumstance that the objected information still remains available after removing the hyperlinks due to this information not appearing on the server of Indymedia, does not remove the unlawful act by Indymedia and cannot therefore take off the previous. The admissibility of hyperlinks is not in general up for general discussion. It points out the contents of the publications to which those hyperlinks offer access.'

Asscher, former member of the IvIR [429], argues in his column [430] that the judgment can be understood but that the argument is imprecise and 'looks lazy'. However, in 2002 the Appeal Court

---

[426] Scientology v XS4ALL et al, 1999. WWW: <http://www.xs4all.nl/uk/news/overview/verd2eng.html>.

[427] Scientology v XS4ALL et al, 2003. The Court of Appeal in The Hague today rejected all of Scientology's claims in appeal in Scientology's action against XS4ALL, Karin Spaink and ten other internet providers. The court concluded that Mrs. Spaink's publications which quoted from works of Scientology were completely legal. In this case, the court said, freedom of opinion does not take second place to enforcement of copyright. The court has found against Scientology on all points. WWW: <http://www.xs4all.nl/uk/news/overview/scientology.pdf>.

In July 2005, a few days before the Supreme Court was to rule upon the case, Scientology dropped its appeal. The Supreme Court accepts the cult's withdrawal, thereby pre-empting the need for a legal evaluation of the case. In November 2005 the Supreme Court decided that the previous ruling, which was in favour of XS4ALL et al, still holds.

428 Voorzieningenrechter Court Amsterdam 25-04-2002 (Deutsche Bahn v XS4all) and Voorzieningenrechter Court Amsterdam 20-06-2002 (Deutsche Bahn v Indymedia). Published in Dutch: Computerrecht 2002-5.LJN-nr. AE4427, r.o 8.

The Amsterdam Appeal Court has 07-11-2002 pronounced judgment in the Radikal case. The Court has dismissed the grounds for appeal presented by XS4ALL and has upheld the judgment delivered in the interim remedies proceedings. WWW: <http://indymedia.nl/en/static/DB/> ; WWW: <http//www.ivir.nl/publicaties/asscher/deutschebahn-xs4all.html>.

[429] Instituut voor Informatierecht (Institute of Information Law, University of Amsterdam); WWW: <http://www.ivir.nl>.

dismissed the grounds for appeal which forced provider XS4ALL to block information on a customer's Website and disclose the customer's name and address.

Wrapping-up the Cases

As hyper-linking is – in technological view – an external representation of the internal linking of software components, the conclusion – by analogy with the DB v Indymedia judgment – can be that it does not matter how a virtual identity is representing itself and from which source the VID is generated (by human or machine), it has to obey the law.
Also the separate declaratory judgment in the 1999 Scientology case grounded the conclusion that a service provider (e.g. the manager of an involved server) must take action against abuse when the provider is aware of this.
Notwithstanding that both decisions came from courts in the Netherlands, the analogy of their consideration can be persuasive for other courts in other countries.

To prevent court cases in areas that attract criminal liability, it is possible for industry to play a self regulatory role. For example, the European ISP Association has a hotline service to tip-off law enforcement agencies that illegal content is in their jurisdiction. Such a hotline could supplement criminal laws regarding the Internet.

The Internet and Sovereignty

In the early days of the 'open' Internet, it was trendy to state that cyberspace was a foreign autonomous world without borders and that notions of national sovereignty were outdated and did not apply to the Internet.

> *Governments of the Industrial World, you weary giants of flesh and steel, I come from Cyberspace, the new home of Mind. On behalf of the future, I ask you of the past to leave us alone. You are not welcome among us. You have no sovereignty where we gather. ...*
>
> *... Cyberspace does not lie within your borders. Do not think that you can build it, as though it were a public construction project. You cannot. It is an act of nature and it grows itself through our collective actions. ...*
>
> *... We believe that from ethics, enlightened self-interest, and the commonweal, our governance will emerge. Our identities may be distributed across many of your jurisdictions. The only law that all our constituent cultures would generally recognise is the Golden Rule. We hope we will be able to build our particular solutions on that basis. But we cannot accept the solutions you are attempting to impose. ...*
>
> Quotes from 'A Declaration of the Independence of Cyberspace' by John Perry Barlow. [431]

In reality, frontiers are deeply embedded in our frame of mind. Far from hidden or disappearing, borders in cyberworld have been more expressed over time.
Even today, however, you can find some revolutionaries whose views reflect the cyber squatters of the American male early adopters: that 'cyberspace will be recognised ... as a sort of independent jurisdiction.' (Basedow & Kono) [432] This is unlikely to be realised as cyberspace is step by step evolving as an integral part of the real world with its increasing web of international conventions and organisations, thanks to the result of diplomacy, terrorism and international collaboration in public policies.

The Internet is a worldwide network with a potential worldwide range. However, the Internet is pseudo compartmented by regions where language or culture is separated. Also, physically the local networks are centric (before being connected to the main network through an Internet exchange point), due to network design and to political and natural special circumstances (like state borders, mountains, rivers and seas).
The international private law legal relation(ship) is coordinated by the International Private Law (IPL). The primary factors of departure / starting points of the IPL in pointing out or acceptance of jurisdiction have a territorial or geographic character. The Internet has an exterritorial effect. The

---

[430] L Asscher „Deutsche Bahn'; WWW: <http://www.ivir.nl/publicaties/asscher/I&I_Deutsche%20Bahn_sabotteert_Internet.pdf>.

[431] Barlow's manifesto 'A Declaration of the Independence of Cyberspace' was posted to the net on 08-02-1996, the day after the Telecom Reform Act of 1996 was signed into law. WWW: <http://www.eff.org/~barlow/Declaration-Final.html>.

[432] J Basedow and T Kono, (eds) *Legal Aspects of Globalisation: Conflict of Laws, Internet, Capital Markets and Insolvency in a Global Economy* (London: Kluwer Law International, 2000), p. 30.

geography between the jurisdiction and the territory of a state is less strong and visible than the 'classic' legal relationship. Without global positioning the sovereignty is adrift.
For a decade, the geographical domain name of a Website or the suffix of an email address was the only guide available for understanding the location of a user. However, in the case of generic top-level domains (such as the anonymous download Website "mp3.com") this was an unreliable guide. [433]
Today, sophisticated techniques can be used to locate an Internet connected party, even when an online pseudonym or virtual identity is used. The geographic allocation of the terminal access point of IP addresses enables us to locate a user, which not only helps investigators to point out the physical location of the computer, but also generates a new palette of location based services, such as popup screens with local (language adapted) offers of accommodation and entertainment.

Law Will Regulate All?
One could say that a virtual world is a kind of outer space, and does not belong to our real world. Some dream about an independent virtual state that exists free from physical nations. Others imagine an illusory country in which they can live their second life as they such as. They all rely on the *'anytime, anyplace, anyway'* slogan.

Cyberworld is *'a graphic representation of data abstracted from banks of every computer in the human system'* (Gibson)[434] and is *'made by Internet and Cell phone'* (Kokswijk)[435].

> ... Our identities have no bodies, so, unlike you, we cannot obtain order by physical coercion. ... Cyberspace consists of transactions, relationships, and thought itself, arrayed such as a standing wave in the web of our communications. Ours is a world that is both everywhere and nowhere, but it is not where bodies live. ...
>
> Quotes from 'A Declaration of the Independence of Cyberspace' by John Perry Barlow. [431]

Despite the position that 'their' cyberworld is not where the bodies live, the reality is that this virtual world – including the virtual identity – only exist (and is imagined) when the computer (network) is switched on, and when there is electricity to operate it. To be 'made by Internet and Cell phone', electricity is a condition for existence of the virtual world. But, following the electric atoms to the source of supply – such as tracing the money to the criminal [436] – the virtual outer space is enabled (fed) by physical energy sources somewhere on our globe (does not matter: solar cell, gas turbine, wind generator, nuclear power station, chemical accumulator, hydroelectric plant, or some other kind of power source). Cyberworld seems to be a Walhalla with a power cord to the feed of Mother Earth. Disconnecting the power or intermitting the connections route in the servers will surely be an effective way to 'order by physical coercion' the cybernetic flower power.

In fact, as long as the virtual world is essentially depending on real world energy and the sock puppeteers behind the avatars (et cetera) that are controlling their virtual identities from electronic connections somewhere on our planet – the virtual world is under the (territorial) jurisdiction of the real world. This conclusion is supported by the observation that only inhabitants of our earth use the cyberworld, that when value is involved this also finally relates to the monetary system on earth, and that the necessary network – even when it is wireless – is traceable to points of presence all over our globe.

Cyberspace is Not Outer Space
The virtual experience is created by electronic memories, networks and computers all over the world. Gibson states: *'Cyberspace is the total interconnectedness of human beings through computers and telecommunication without regard to physical geography.'* The principle is that the telecom network is interconnected to places all over the world, the virtual world itself is an international topic. When the individual connection to the computer network is started from some computer in a particular geographic area, belonging to the territory of a specific country, one could argue that all content through that individual connection meets the jurisdiction of that country.

---

[433] Top-level domains such as .com, .edu, .gov, .net, .org, .mil etc.

[434] W Gibson, *Neuromancer*. Cite: ' A consensual hallucination experienced daily by billions of legitimate operators, in every nation, by children being taught mathematical concepts ... A graphic representation of data abstracted from banks of every computer in the human system... Lines of light ranged in the nonspace of the mind, clusters and constellations of data. ...' (Ace 1984; Reprint edition 1986) p 51.

[435] J van Kokswijk, *Architectuur van een cybercultuur* (Bergboek, Zwolle 2003) Hypothesis 1.

[436] J Gillespie, 'Follow the Money: Tracing Terrorist Assets' Seminar on International Finance, Harvard Law School (2002). WWW: <http://www.law.harvard.edu/programs/pifs/pdfs/james_gillespie.pdf>.

As digital worlds become increasingly powerful and lifelike, people will employ them for countless real-world purposes, including commerce, education, medicine, law enforcement, and military training. Inevitably, real-world law will regulate them (Balkin) [437]; (Goldsmith & Wu) [438]. It can be concluded that the legal position of the virtual world(s) – being conditionally connected to and by under the (territorial) jurisdiction of the real world on planet Earth – cannot be outer space, nor will it be singly local, regional, national or global. It will be regulated by all.

Can/Should the Internet be Regulated for Identities?

This subchapter discusses if the Internet can be regulated for virtual identities. It also concerns the ways code can control the individuals and groups in a virtual society, and highlights the very close relation between commerce and code.

The Need and Distaste for Rules

New Media, Renewed Law

When a new medium evolves, it seems to be that it's discussed regarding improved regulating more than mediums were in the past. That happened 150 years ago with the telegraph, 50 years ago with the radio, and for 25 years with the premium telephone numbers and other illicit businesses. The regulators need to learn from the lessons of the telegraph and telephone before they make the same mistakes regarding the Internet. For example, the regulators' banning of 1-900 numbers in the United States has only established an off-shore gold mine for some enterprising people. Since these calls have been routed through Venezuela or Guyana, the US does not receive any tax revenue or benefit from more employment. The same has been true for gambling, which has reaped the benefits of off-shore accounts and the speed of the Internet. The ability to hide in other jurisdictions and through anonymity makes an exit particularly easy in cyberspace and this is why governments are unable to effectively impose their laws in cyberspace. Nonetheless, such an exit does reduce opportunities for wealth creation within the jurisdictions that are trying to control the various activities.

Licence to Chill

The conservatives seem to like the morality of censoring the phone numbers we can call, the Internet sites we can explore, the movies we can watch, and the activities we can choose to do in our free time. However, in the end, we will pay a higher price for basic phone service, for our Internet service provider, and all other suppliers. In addition, the governments forfeit millions of euros in tax revenue. So, regulation of the Internet is open to question.

> '*A new and complex regulation requires all Internet users to be licensed by the Ministry of Culture and Information before posting information on a Website. The provision of information that is inconsistent with the license could lead to the licence's removal, a fine or even imprisonment. Foreign entities such as news agencies, embassies and international or non-governmental organisations would have to receive an additional permit from the Ministry of Foreign Affairs.*' ISPO Trends [439]

In 2000 US President Clinton pointed out the opposite:

> '*What we're going to try to do today is to talk about what the government's responsibility is for our own systems and networks, what the private sector's responsibility is -- and, as I said before, how to talk about having adequate security, how to protect privacy and civil liberties but also how to keep the Internet open. Keep in mind that one of the reasons this thing has worked so well is it has been free of government regulation.*' Bill Clinton [440]

The Internet can, should and will be subject to social order. There will be rules, often called regulations. The question is how do we set those rules? And how do we get the values that the Internet will maximise in our society? One of the difficulties we have is the minute we say 'we need

---

[437] JM Balkin and BS Novick (ed), *The State of Play: Law, Games, and Virtual Worlds (Ex Machina: Law, Technology, and Society)* (New York University Press, 2006).

[438] J Goldsmith and T Wu, *Who controls the Internet? Illusions of a borderless world* (Oxford Press 2006).

[439] ISPO Trends, 18-06-1997; WWW: <http://europa.eu.int/ISPO/services/i_trends.html>.

[440] WJB Clinton, 15-02-2000; WWW: <http://www.pbs.org/newshour/bb/cyberspace/jan-june00/internet_security_2-15.html>.

a rule here', everyone says 'don't regulate the Internet'. Only after we recognise that we need some rules for cyberspace, just like we have in real space, then we can start to address the problem in an open, educated and informed way.

Benson argues that the Internet needs to be regulated and that it will be by the private sector, but also by the public sector. [441] He points out that it is possible that a group may voluntarily cooperate (as opposed to being coerced into cooperating) to enforce a law where no identifiable victim exists if virtually everyone in the relevant group believes that the law should be enforced. But in a stateless system of law and enforcement, the allocation of enforcement resources would be determined by individual willingness to pay rather than by political strength or bureaucratic discretion over common pool resources. [442]

Friedman once argued that people who want to control other people's lives are rarely eager to pay for the privilege. [443] They usually expect to be paid for the service they provide for their victims. A private system of law would clearly be strongly biased toward individual freedom when individual action does no harm to another's physical person or property. As in the real world the cyberworld is full of communities. The possibility of a community having its own laws, differing substantially from other communities, does not mean that an irrational patchwork of entirely different law systems will exist.

History demonstrates that standardisation of many aspects of customary law over very large geographic areas would arise. Benson shows that there certainly may be relatively minor differences, but perhaps even less differentiation would occur than exists from state to state, and even from city to city, under the political system of law we currently have. [444] Consider the privately developed English language, for example.

Tremendous levels of standardisation dominate all the regional differences in language, and in customary law. Benson argues that a system that emphasises individual responsibility and liberty can be established under customary law with private sector institutions for enforcement and adjudication. Such a system may not be perfect: e.g., some free riding may occur, states Benson in 1990. Therefore some may suggest that limited government involvement – doing those few things that it might do better than markets – would be superior to complete privatisation.

Friedman answered this question in the following way:

> 'Perhaps it would be - if the government stayed that way. . . . One cannot simply build any imaginable characteristics into a government; governments have their own internal dynamic. And the internal dynamic of limited governments is something with which we, to our sorrow, have a good deal of practical experience . . . the logic of limited government is to grow. There are obvious reasons for that in the nature of government, and plenty of evidence. Constitutions provide, at the most, a modest and temporary restraint. As Murray Rothbard is supposed to have said, the idea of a limited government that stays limited is truly Utopian.' [445]

Every aspect of government involvement in law and order started out to be very limited (or nonexistent). Contrary to a wide-spread misconception private action has been the predominating mode of law enforcement throughout most of human history. Pervasive government involvement in law enforcement is a recent phenomenon.

> 'Royal courts in England initially had very limited jurisdictions. Then they began competing with other courts in adjudicating increasingly more diverse laws. They had a 'competitive' advantage in that part of the cost of using them was not born by litigants. Various interest groups were happy to shift their costs for protection services and the enforcement of their laws onto others by using government courts, and later government watchmen, police, prosecutors, and so on. Government entities were happy to oblige. The combination of power seeking and bureaucratic growth by government officials

---

[441] BL Benson, 'The Spontaneous Evolution of Cyber Law; Norms, Property Rights, Contracting, Dispute Resolution and Enforcement Without the State' (2005) WWW: <http://www.sjsu.edu/depts/economics/faculty/powell/docs/econ206/Cyber-Law-Evolution.pdf>

[442] For discussions of various aspects of government law enforcement as common pool resources, see:

BL Benson, 'Corruption in Law Enforcement: One Consequence of 'The Tragedy of the Commons' Arising With Public Allocation Processes', International Review of Law and Economics 8 (June 1988) pp 73-84;

BL Benson and LA Jr Wollan, 'Prison Crowding and Judicial Incentives', Madison Paper Series, No. 3 (May 1989) pp 1-21;

R Neely, *Why Courts Don't Work* (New York: McGraw-Hill, 1982).

[443] DD Friedman, *The Machinery of Freedom: Guide to a Radical Capitalism* (New York: Harper and Row, 1973). p. 173-174.

[444] BL Benson, 'The Spontaneous Evolution of Commercial Law', Southern Economic Journal, V 55, No 3 (Jan 1989) pp 644-61;

BL Benson, 'The Enterprise of Law: Justice Without the State' (Pacific Research Institute 1990).

[445] DD Friedman, *The Machinery of Freedom: Guide to Radical Capitalism*, (New York: Harper and Row, 1973) p. 200-201.

and transfer (or rent) seeking by interest groups inevitably turns limited government into big government.'
Benson [446]

It is well known that for several centuries *lex mercatoria* developed mostly outside the sphere of public legislation and law enforcement, it was enforced by private courts, reputational concerns and ostracism.
The survey of Schönfelder shows that these (semi)private courts were the historical predecessors of modern arbitration courts. [447] 'The adoption of the rules of *lex mercatoria* by state courts was a gradual process which extended over centuries. Merchants who failed to comply with the rulings of merchant courts found themselves cut off from valuable business opportunities because honourable traders often learned about their misbehaviour and refused to deal with them. Arbitration has been a wide-spread means of dispute settlement even though in most nations it was only in the second half of the nineteenth century that arbitral awards started to be enforced by state courts.'

Control without Government

Friedman describes how we, without government, could settle the disputes that are now settled in courts of law but how could we protect ourselves from criminals? [448] He presents the fictitious case of Tannahelp, Dawn Defense, the automatic camera's evidence and the stolen television set. This example refers to the resolution of disputes involving contracts between well-established firms. A large fraction of such disputes are now settled not by government courts but by private arbitration. The firms, when they draw up a contract, specify a procedure for arbitrating any dispute that may arise. Thus they avoid the expense and delay of the courts. The involved arbitrator has no police force. His function is to render decisions, not to enforce them. Currently, arbitrated decisions are usually enforceable in the government courts, but that is a recent development; historically, enforcement came from a firm's desire to maintain its reputation. Before labelling a society in which different people are under different laws chaotic and unjust, remember that in our society the law under which you are judged depends on the country, state, and even city in which you happen to be. Under the arrangements I am describing, it depends instead on your protective agency and the agency of the person you accuse of a crime or who accuses you of a crime.

In such a society law is produced on the market. A court supports itself by charging for the service of arbitrating disputes. Its success depends on its reputation for honesty, reliability, and promptness and on the desirability to potential customers of the particular set of laws it judges by. The immediate customers are protection agencies. But the protection agency is itself selling a product to its customers. Part of that product is the legal system, or systems, of the courts it patronises and under which its customers will consequently be judged. Each protection agency will try to patronise those courts under whose legal system its customers would like to live. Friedman points out the objections that may be raised to such free-market courts.
While many people may believe that the state must step in to provide laws for cyberspace, the fact is that geographically defined nation-states are actually not capable of establishing effective law in cyberspace. Rustad suggests three of the primary reasons for this: (1) jurisdictional issues, (2) anonymity in cyberspace, and (3) the high opportunity cost of law enforcement resources devoted to cyber detection and prosecution. [449] Jurisdictional problems are a major impediment to nation-states' efforts to govern in cyberspace:

> *'Cyberspace radically undermines the relationship between legally significant (online) phenomena and physical location. The rise of the global computer network is destroying the link between geographical location and: (1) the power of local governments to assert control over online behavior; (2) the effects of online behavior on individuals or things; (3) the legitimacy of local sovereign's efforts to regulate global phenomena; and (4) the ability of physical location to give notice of which set of rules apply. The Net thus radically subverts the system of rule-making based on borders between physical spaces, at least with respect to the claim that Cyberspace should naturally be governed by territorially defined rules'*
> Johnson and Post [450]

---

[446] BL Benson, 'The Enterprise of Law: Justice Without the State' (Pacific Research Institute, 1990).

[447] B Schönfelder, 'The Puzzling Underuse of Arbitration in Post-Communism – A Law and Economics Analysis' Freiberg Working Papers (2005). WWW: <http://www.tu-freiberg.de/~wwwfak6/files/paper/schoenfe_7_2005.pdf>.

[448] DD Friedman, *Police, Courts, and Laws – on the Market*; in: The Machinery of Freedom: Guide to Radical Capitalism (Open Court, 1973; 1989 second edition).

[449] ML Rustad, 'Private Enforcement of Cybercrime on the Electronic Frontier,' *Southern California Interdisciplinary Law Journal*, 11 (2001) pp 85-86 in pp 63-116.

[450] DR Johnson and D Post, 'Law and Borders - The Rise of Law in Cyberspace,' *Stanford Law Review* 48 (1996) p 1370 in pp 1367-1402.

Due to the focus on jurisdictional problems very few law enforcement tools for the Internet are developed. Over viewing today's enforcement strategies (table 2) it is clear that both spontaneous and executed enforcement strategies are lacking effectual and sufficient cyber equivalents with the exception of using software code as control.

Enforcement Strategies	Spontaneous Enforcement		Executed Enforcement	
Strategy	Convince	Cooperation	Control	Sanction
Primary intended outcome:	Norm reinforcement	Norm reinforcement	Deterrence	Deterrence
Core value:	Legitimacy (fairness)	Social following and fear for reputation loss	Visible authoritative control	Penalty, fine
Better compliance by:	Appeal to sense of duty	Group loyalty and self-interest. Web broadening of parties concerned	Formal supervision and surveillance	Punitive enforcement

Table 2: Overview enforcement strategies, based on The Dutch Centre of Expertise in Legal Enforcement

Projecting these strategies to the virtual environment it is becoming clear that both control and sanction are different. Only when 'visible' annex 'formal supervision and surveillance' are transformed to 'being monitored by electronic equipment' and 'penalty' and 'punitive enforcement' are translated in 'limitation by software code' we could use these strategies in cyberspace. That declares the drive of enforcers to take extreme actions of monitoring as much as possible the information channels.

Control by Code

The Palette of Codes

Regulating by software code is one of the emerging ways of controlling the Internet and also the applications that use various forms of information technology. Very popular today are the logging and mining of all assembled data to file on potential useful information 'just in case'. Second common practice is the use of automatic software ('bots') and other various types of communication control, e.g. limiting the features in ability (like usage time, option, level, volume), and/or booting [451], blocking [452] and banning the IP address of a connected computer. This kind of software is executing actions more or less autonomously to an increasing degree, like 'intelligent agents' already do in advanced software [453].

Code to Rule the Citizen

Governments are introducing, managing and using digitised personal identification and authentication systems in their service relationships with citizens in addition to, and increasingly in replacement of, traditional forms of personal identification and authentication. Digitised personal identification systems can offer customer convenience; citizen mobility and empowerment; efficiency and/or effectiveness of public service provision, including joined-up government; and the enhancement of public safety and security, including general law enforcement.

Lips, Taylor & Organ point out that these systems therefore not only appear to enable the modernisation of government; they also enable government to fulfil its service providing functions, through its ability to authenticate personal identifiers provided by citizens in e-government relationships. [454] Authentication, or the assurance that a person is who (s)he says (s)he is, is

---

[451] Act of ejecting a user from a chat room. WWW: <http://en.wikipedia.org/wiki/Booting_(computer_slang)>.

[452] Expulsion. WWW: <http://en.wikipedia.org/wiki/IP_Ban> and WWW: <http://en.wikipedia.org/wiki/IP_banning>.
Blocking is also used for vandal attack contributions to Wikipedia; see the interesting debate about over 1.000 edits by the US Congress; WWW: <http://en.wikinews.org/wiki/Talk:Congressional_staff_actions_prompt_Wikipedia_investigation>; WWW: <http://en.wikinews.org/wiki/Wikinews_investigates_Wikipedia_usage_by_U.S._Senate_staff_members>, and WWW: <http://en.wikipedia.org/wiki/Wikipedia:Requests_for_comment/ United_States_Congress>.
WWW: <http://en.wikinews.org/wiki/Talk:Congressional_staff_actions_prompt_Wikipedia_investigation>.

[453] Intelligent Agent (IA) as assisting software agent WWW: <http://en.wikipedia.org/wiki/Intelligent_agent>.

[454] M Lips, JA Taylor, and J Organ, *Identity Management as Public Innovation: Looking Beyond ID Cards and Authentication systems*. In: Bekkers, Van Duivenboden & Thaens (eds), 'ICT and Public Innovation: Assessing the Modernisation of Public Administration' (IOS Press, Amsterdam 2006).

generally acknowledged as an essential requirement for the provision of many government services to citizens. They argue that digitised personal identification and authentication systems thereby become the *sine qua non* of successful e-government. With these new digital forms of personal identification, authentication and identity management we seem to have arrived into a new *révolution identificatoire* in the public domain, where a law of informational identity, a so-called *ius informationis*, may soon replace the existing models of citizenship attribution in the analogue world, *ius soli* and *ius sanguinis*.

Code is Control

Government and Commerce are changing the net, from an unregulable place to a highly regulable place, argues Lessig [455]. The most important Web 2.0 concept is that the infrastructure of the Internet will become increasingly controlled and regulable through digital identity technologies. Only stopping this control push will result in having an Internet closer to the values of its original design. Lessig explained in Code v2.0 (2006) the differences in networks of cyber places: *Harvard vs. Chicago* (pp 33-35). The networks thus differ in the extent to which they make behaviour within each network regulable. This difference is simply a matter of code – a difference in the software and hardware that grants users access. Different code makes differently regulable networks. So called regulability is thus a function of design.

Following Lessig's view these two networks are just two points on a spectrum of possible network designs. 'At one extreme we might place the Internet — a network defined by a suite of protocols that are open and non proprietary and that require no personal identification to be accessed and used. At the other extreme are traditional closed, proprietary networks, which grant access only to those with express authorisation; control, therefore, is tight. In between are networks that mix elements of both. These mixed networks add a layer of control to the otherwise uncontrolled Internet. They layer elements of control on top.' The practice of today is that new professional networks are set up in a full regulable architecture.

Such as Plato's penal code in the past (that regulates the penalty after being publicly discussed), transparent regulation in cyberspace can help us see something important about how all this works. Regulability will also introduce a regulator ('code') whose significance we don't yet fully understand. 'Regulation by Code' will render ambiguous certain values that are fundamental to our tradition.

Lessig argues that a 'latent ambiguity' will require us to make a choice among many sovereigns. In the end the hardest problem will be to reckon these 'competing sovereigns,' as they each act to mark this space with their own distinctive values.
Ten years ago we all thought that the government(s) could never regulate the Net, which was a good thing. Today, attitudes are different. In the US there is still the commonplace (supported by Perritt) [456] that government can't regulate and has to back private solutions, but in a world drowning in spam, computer viruses, identity theft, and sexual extremes, their resolve against regulation has weakened.

Despite this expanding framework, Lloyd suggests that 'just as the industrial revolution rendered obsolete aspects of law based on notions of an agrarian society, a legal system focusing on issues of ownership, control and use of physical objects must reorient itself to suit the requirements of an information society.' [457]

More generally, Lessig argues that there are actually four major regulators – Law, Norms, Market, and Architecture – each of which has a profound impact on society and whose implications must be considered. Following Plato's theory that there are at least two kinds of code, legal and social, there also can be a technical code that enables software to enforce strict compliance with the law. Cyberspace not only changes the technology of copying but also the power of the law to protect against illegal copying, even when there is fair use behaviour. The call 'anytime anywhere' sounds differently when control can be anywhere and anytime, from things to humans.

> In his revised book 'Code And Other Laws of Cyberspace v2.0' [458] Lessig resurrects the prophesying of two science-fiction writers who told stories about cyberspace's future on a congress in 1996: [459]

---

[455] L Lessig, *Code and other laws of cyberspace*. (Basic books, 1999).

[456] HH Perritt, 'Towards a Hybrid Regulatory Scheme for the Internet'. University of Chicago Legal Forum 215 (2001).

[457] IJ Lloyd, *Information Technology Law*, (Kluwer 2004) p 6.

[458] L Lessig, *Code And Other Laws of Cyberspace, v2.0* (Basic books 2006) WWW: <http://pdf.codev2.cc/Lessig-Codev2.pdf>.

[459] 1996, annual conference 'Computers, Freedom, and Privacy' (CFP).

'Vernor Vinge spoke about 'ubiquitous law enforcement,' made possible by 'fine-grained distributed systems'; through computer chips linked by the Net to every part of social life, a portion would be dedicated to the government's use. As this network of control became woven into every part of social life, it would be just a matter of time, Vinge threatened, before government claimed its fair share of control. Each new generation of code would increase this power of government. The future would be a world of perfect regulation, and the architecture of distributed computing – the Internet and its attachments – would make that possible.

Tom Maddox followed Vinge. His vision was very similar, though the source of control different. The government's power would not come just from chips. The real source of power, Maddox argues, was an alliance between government and commerce. Commerce, such as government, fares better in a better regulated world. Property is more secure, data are more easily captured, and disruption is less of a risk. The future would be a pact between these two forces of social order. When these two authors spoke, the future they described was not yet present. Cyberspace was increasingly everywhere, but it was hard to imagine it tamed to serve the ends of government. And commerce was certainly interested, though credit card companies were still warning customers to stay far away from the Net. In 1995 The Net was an exploding social space of something, but hardly an exploding space of social control. Day by day the Internet is becoming the perfect space of regulation and you can imagine how commerce would play a role in this regulation. In many situations control will be coded, by commerce, with the backing of governments.

Vinge and Maddox were first-generation theorists of cyberspace. They could tell their stories about perfect control because they lived in a world that couldn't be controlled. They could connect with their audience because it too wanted to resist the future they described. Envisioning this impossible world was a sport. Now the impossible has been made real. Much of the control in Vinge and Maddox's stories that struck many of their listeners as Orwellian now seems quite reasonable. It is possible to imagine the system of perfect regulation that Vinge described. (...) It is inevitable that an increasingly large part of the Internet will be fed by commerce, and most don't see anything wrong with that either. Indeed, we live in a time (again) when it is commonplace to say: let business take care of things. Let business self-regulate the Net. Net commerce is now the hero. The future is Vinge and Maddox's accounts together, not either alone. If we were only in for the dystopia described by Vinge, we as a culture would have an obvious and powerful response: Orwell gave us the tools, and Stalin gave us the resolve to resist the totalitarian state. Is a spying and invasive Net, controlled by governments, our future? 1984 is in our past.'

Code and Commerce

Lessig argues that we live life in real space, subject to the effects of code. We live ordinary lives, subject to the effects of code. We live social and political lives, subject to the effects of code. Code regulates all these aspects of our lives, more pervasively over time than any other regulator in our life. Should we remain passive about this regulator? Should we let it affect us without doing anything in return?

Governments should intervene, at a minimum, when a private action has public consequences; when short-sighted actions threaten to cause long-term harm; when failure to intervene undermines significant constitutional values and important individual rights; and when a form of life emerges that may threaten values we believe to be fundamental.

Lessig advocates a different response: we need to think collectively and sensibly about how this emerging reality will affect our lives. Doing nothing is not an answer; something can and should be done. As Lessig states in one of his speeches: 'Creativity and innovation always build on the past. The past always tries to control the creativity that builds on it. Free societies enable the future by limiting the past. Ours is less and less a free society.' As I argued before: each new medium gets more control than its predecessors.

Law as Country Code

Law is no (self-reliant autonomous) entity on its own. Ultimately the 'law' is laid down by what governments in their country decide, enabled by their constitution. The ideal vision of a population can be achieved by power, but normally, in civilised countries, law has a social character and is a legal code parallel with the social code. Only in exceptional situations can it be temporarily acceptable that power is law, but once the incident is over it is not only 'business as usual' but also 'law as usual'.
Even when more countries share their social ideals, and like to strengthen these ideals by conventions, there are constitutional or political restraints as a result of which – even modern

civilised – countries do not accept and enforce some international conventions, pacts and treaties, such as Norway, Japan, Iceland in the case of the whale hunting (Washington convention) [460], and Australia and the USA in the air pollution case (Kyoto convention) [461].
Also there are countries that protest – even when they are officially united in a higher governmental entity (such as Europe) – against some 'higher' laws that seem to lose their identity, such as France in the case of the European constitution. Also, the English are of a different breed as they guard their identity and autonomy by for instance driving on the left and by using their currency in Pounds instead of the common Euro. In many cases there is a mismatch between the political ideals and the economic and/or military reality. (For example India and USA) [462]

Cyberspace is not much better regulated. Because of US diplomatic pressure the United Nations have temporised over long discussions concerning the governance of the Internet. Finally since 2004 they care about the government of the Internet but still without taking decisive measures to counter Internet governance. In 2005 the UN's Working Group Internet Governance (WGIG) recommended e.g.: [463]

73. **Regional and national coordination**

The WGIG noted that international coordination needs to build on policy coordination at the national level. Global Internet governance can only be effective if there is coherence with regional, subregional and national-level policies. The WGIG therefore recommends:

(a) That the multi-stakeholder approach be implemented as far as possible in all regions in order for the work on Internet governance to be fully supported at the regional and subregional levels;

(b) That coordination be established among all stakeholders at the national level and a multi-stakeholder national Internet governance steering committee or similar body be set up.

79. **Internet stability, security and cybercrime**

• Efforts should be made, in conjunction with all stakeholders, to create arrangements and procedures between national law enforcement agencies consistent with the appropriate protection of privacy, personal data and other human rights.

• Governments, in cooperation with all stakeholders, should explore and develop tools and mechanisms, including treaties and cooperation, to allow for effective criminal investigation and prosecution of crimes committed in cyberspace and against networks and technological resources, addressing the problem of cross-border jurisdiction, regardless of the territory from which the crime was committed and/or the location of the technological means used, while respecting sovereignty.

83. **Data protection and privacy rights**

• Encourage countries that lack privacy and/or personal data-protection legislation to develop clear rules and legal frameworks, with the participation of all stakeholders, to protect citizens against the misuse of personal data, particularly countries with no legal tradition in these fields.

• The broad set of privacy-related issues described in the Background Report be discussed in a multi-stakeholder setting so as to define practices to address them.

---

[460] International Convention for the Regulation of Whaling (1946); The International Whaling Commission (IWC); WWW: <http://www.iwcoffice.org/>.

[461] The Kyoto Protocol to the United Nations Framework Convention on Climate Change is an amendment to the international treaty on climate change, assigning mandatory targets for reduction of greenhouse gas emissions to signatory nations. Kyoto, 18-05-1973. Kyoto Amending Supplement No. 13/1993. WWW: <http://www.unece.org/trade/kyoto/>. Revised Kyoto convention: WWW: <http//www.wcoomd.org/ie/en/Topics_Issues/ FacilitationCustomsProcedures/kyoto/kyreport.html>.

[462] Observing the attitude of some governments with respect to international treaties their record is abysmal. For instance the US Government has torn up the following significant international treaties:
- The Anti-Ballistic Missile treaty (start of development of a missile defence system);
- The Biological Weapons Convention (veto verification protocol);
- Cooperative Threat Reduction (cut back aid to Russia to destroy weapons);
- START II and III (the cut in warheads is off);
- The Chemical Weapons Convention (USA wants to be able to refuse inspections);
- The Nuclear Non-Proliferation treaty;
- The Comprehensive Test-Ban treaty;
- The Landmine Treaty.

Source: WWW: <http//www.euronet.nl/users/e_wesker/USpol.html>.

[463] WJ Drake (ed), 'Reforming Internet Governance; Perspectives from the Working group on Internet Governance (WGIG)' (2005); WWW: <http//www.wgig.org/docs/book/WGIG_book.pdf>.

- The policies governing the WHOIS databases should be revised to take into account the existence of applicable privacy legislation in the countries of the registrar and of the registrant.

- Policy and privacy requirements for global electronic authentication systems should be defined in a multi-stakeholder setting; efforts should then be made to develop open technical proposals for electronic authentication that meet such requirements.

A report full of 'be's, need's and should's' will not hasten the worldwide governments to come on to one track regarding the effective global governance of the Internet. Also, the Internet Governance Forum, as proposed in the WGIG Report, would not be a self-regulatory body because it would not have any enforcement powers. The forum would be a gathering platform to exchange views and share best practices. In the discussions one government after another took the floor to express it support for the creation of a Forum, albeit with slightly different formulations of its purpose and potential functions. Industry participants regrouped to a position that the Forum would need to be well managed, low cost, clearly tasked, and so on.

Finally in 2006 the 1st Meeting of the Internet Governance Forum (IGF) [464] took place in Greece. At this inaugural meeting, the issues were divided into four main categories: openness, security, diversity and access. It seems to be the first international organisation that invites the global citizen to interact online by instant blog, in live chats or discussions, to participate in polls and to help edit the IGFs wiki pages.

Summary Chapter 7: 'Legal Issues and Policy of Virtual Identity I'

The progress of order and law started in ancient times. Originally, law had a non-territorial basis. The principle of territorial sovereignty was unknown in the ancient world. It seems that the Peace of Westphalia in 1648 was a decisive year, as the idea of exclusive territorial sovereignty replaced the theory of the personality of laws as the fundamental principle of international intercourse. In the paragraph Sovereignty and Jurisdiction different cases about cross-border law and jurisdiction are presented.

The Internet and sovereignty is discussed in order to make it clear that cyberspace is not outer space as some prophesied in the 1990s. The question can/should the Internet be regulated for identities is fascinating. But in the meantime, in daily life, bits by bites, the Internet is regulated by all kinds of governmental law, by jurisprudence, and by a layer of non-transparent software code. The regulating by computer source code is one of the emerging ways of controlling the Internet; and also the - mostly commercial - applications use various forms of information technology to control the individual behaviour in the virtual society. Code and commerce meet, blessed by the national governments. When start the case of *Legal Insecurity v User License Agreement* ?

Each new medium seems to be discussed regarding regulation. That happened with the printing machine, the telegraph, the telephone, the radio, the television, the cell phone, and other illicit businesses. Examining intently the historical cases it can be learned from these lessons, of for instance the telegraph development, that the same mistakes regarding regulation are made over and over again. Meanwhile, enforcement meets a gap between real and cyber. The comparison of the physical with the virtual societies in the way they are controlled by the elements of a comprehensive system of social control can be found in the forthcoming chapter.

---

[464] Internet Governance Forum (IGF). WWW: <http://www.intgovforum.org/> and WWW: <http://igf2006.info/>.

8   Legal Issues and Policy of Virtual Identity   Part II

To predict the behavior of ordinary people in advance, you only have to assume that they will always try to escape a disagreeable situation with the smallest possible expenditure of intelligence.

Friedrich Nietzsche [465]

Introduction

With human behaviour in a virtual environment the legal issues of law and persons and law and property come under discussion. The basis of intellectual property law for instance is not only a-territorial, but is also related to the economy. Three variables construct the basis of law and economic analysis: communities, time and individuals. The legally intangible anonymous identity in the cyberworld shows the position of the virtual environment in the real world system of rules. The use of virtual identity as a cover for criminal conduct is spotted, similarly, as is the idea of intangible anonymity.

Real Behaviour in Virtual Environments

The role of the computer-as-environment is to mediate and support interaction among multiple humans. In virtual environments, information technology generates, provides, and captures rich and natural sensory signals to and from the human. Varieties of embedded computing or – so called – augmented virtual reality systems mix together aspects of the material and virtual. One of the related aspects is law. Broadly viewed, this concerns the real behaviour of persons in virtual environments.

Behavioural Control in Virtual Society

Social Control

In the comparison between the physical and virtual societies it is clear that the virtual society has adopted most elements of the physical society but has extended them with the software code control elements, as it is much more effective to exert control by making proscribed actions difficult or dangerous in the first place. This can be accomplished – in the digital environment – by changing the software code so that the proscribed behaviour cannot occur, or by monitoring behaviour in such a way that it is a trivial matter to find the perpetrator. In this manner, the code of law is applied to force the code of software into the service of social control. This kind of software control can be used for rules and sanctions such as Ellickson points out in his 'elements of a comprehensive system of social control' [466].

Important to realise in this context is what was stated in chapter 2: with the discovery of information-technology, and especially the 'virtual worlds', all regular natural limits and 'laws' (about time and space) were passed, so human behaviour has become the only thing left to be 'restricted'. Restricting yourself in behaviour, so called 'self control', is supported in the virtual society at the first party level by automatic software ('bots') and the second party control is supported by various types of communication control, e.g. limiting the features in ability (like usage time, option, level, volume), and/or booting [467], blocking [468] and banning the IP address of a connected computer.

---

[465] F Nietzsche (1844–1900), *Human, All-Too-Human*, 'Man Alone With Himself,' aphorism 551, 'The Prophet's Trick' (1878) in: *Sämtliche Werke: Kritische Studienausgabe*, vol. 2, p. 330, eds. G Colli and M Montinari (De Gruyter, Berlin 1980).

[466] RC Ellickson, *Order without Law: How Neighbors Settle Disputes* (Harvard University Press, Cambridge 1991). See also the review of: David Friedman, 'Less Law Than Meets the Eye' in: *University of Michigan Law Review* vol. 90 no. 6, (May 1992) pp 1444-1452. WWW: <http://www.daviddfriedman.com/Academic/Less_Law/Less_Law.html>.

[467] Act of ejecting a user from a chat room; WWW: <http://en.wikipedia.org/wiki/Booting_(computer_slang)>.

Comparing the physical with the virtual societies in the way they are controlled (especially to the topic 'virtual identity') the elements of a comprehensive system of social control can be conveniently arranged and exposed in the table 3. This table shows that 'rules' and 'sanction' in the virtual society are supported by the enablers of cyberspace: electronic equipment. Human behaviour is strongly directed by – mostly non-transparent – software code, sometimes in an abrupt inhuman way.

Virtual Identity Controller	Direction of Human Behaviour in Physical Society			Direction of Human Behaviour in Virtual Society		
	Rules	Sanction	Combined System	Rules	Sanction	Combined System
First-party control						
Actor	Personal ethics & values	Self-sanction	Self-control	Personal ethics & Netiquette	Self-sanction + Software code (limiting by bot)	Self-control + Software code (limiting by bot)
Second-party control						
Person Acted Upon	Contracts	Personal Self-help	Promisee-enforced contracts	User license; Reminders by 'pop ups' in software code	Correction and Ejection by administrators/ moderators or/ and software	Software code (limit of options etc., booting, IP-Blocking, or IP-Banning)
Third-party control						
Social Forces	Norms	Self-help by vicar et al	Informal control	Situational & peer groups norms	Shaming & Blaming; Expulsion	VID-ignoring, (by software code) & Stress by peer leader
Organisation	Organisational rules	Organisational enforcement	Organisational control	Organisational rules + in software code	Organisational enforcement by software code	Organisational visible strain + track & trace
Government	Law	State enforcement	Legal system	Some law & Jurisprudence	State enforcement (if possible)	Legal system+ Software code (limiting by bot)
International boards	Treaty	Cooperative enforcement	Legal system			

Table 3: Elements of a comprehensive system of social control [469] (© Van Kokswijk, 2007)

Significant is the broad acceptance of software code as a combined system of rules and sanctions in the virtual society. The result is a movement toward increasingly secure private property rights under 'customary law'. Indeed, as Ellickson writes, 'There is abundant evidence that a ... group need not make a conscious decision to establish private property rights. ... People who repeatedly interact can generate institutions through communication, monitoring, and sanctioning.' [470]

Therefore, no central authority with coercive powers is necessary to produce law in such a cooperative social order. Benson argues that coercion is only required when there are strong incentives to resist, generally because the law grossly discriminates between individuals or groups in the allocation of rights and wealth. [471] The same situation seems to happen on the Internet, where anonymous moderators and operators often rule the virtual environment. When comparing the standards in different social communities the control various between dictating and deliberating. [472]

---

[468] Expulsion; WWW: <http://en.wikipedia.org/wiki/IP_Ban> , and WWW: <http://en.wikipedia.org/wiki/IP_banning> Blocking is also used for vandal attack contributions to Wikipedia; see the interesting debate about over 1.000 edits by the US Congress: WWW: <http://en.wikipedia.org/wiki/Talk:Congressional_staff_actions_prompt_Wikipedia_ investigation> ; WWW: <http://en.wikinews.org/wiki/Wikinews_investigates_Wikipedia_usage_by_U.S._ Senate_staff_members> and WWW: <http://en.wikipedia.org/wiki/Wikipedia:Requests_for_comment/United_States_Congress>.

[469] Physical society input adapted from Ellickson in: RC Ellickson, *Order without Law: How Neighbors Settle Disputes* (Harvard University Press, Cambridge 1991).

[470] RC Ellickson, 'Property in Land', *Yale Law Journal*, vol. 102 (1993): 1366.

[471] BL Benson, 'Where Does Law Come From?' (1997) WWW: <http://www.independent.org/publications/article.asp?id=202>.

[472] In 'the old' IRC an operator could only be installed as operator when some others accepted her/him as operator. Some operators were pure dictators, and keep on charge by staying full time online, like Badass, ^___^ , an IRC operator who finds it

Status of Virtual Identity in Law of Persons and Property

In this subtitle the position of virtual identity in various laws is observed. The real anonymity of a virtual identity (VID) is discussed, same as the legal intangibility.

Hidden Helpers

Since the Internet virtual identities have step by step become interwoven in the (trans)actions of persons but neglected in the laws regarding persons and property. Most people don't realise that they are using VIDs when entering a marketplace (such as eBay), a chat room, or a dating site. They take for granted that an artificial identity is representing them in selecting, bidding, or dealing during their absence. They mostly are unaware that a lot of electronic transactions have already been completed by 'intelligent agents' (IA), even when the IA has not presented itself to them as an artificial (virtual) identity.

On the other hand, Bogdanowivz and Beslay argue that it is generally known that opportunities to create fictitious virtual identities, potentially in fully fictitious environments are highly exploited in a digital context. [473] They say that the creation of virtual identities can be motivated by privacy and security concerns, convenience or leisure. Virtual identities may disappear without leaving traces. Consequently, the concept of these virtual identities is in contradiction with criteria of permanency and physical reality that are expected in any identification process linking a physical individual to a set of digital data. Consequently the use of multiple 'virtual identities' will have to be regulated in the law regarding persons.

VID in Law of Persons

The VID is recognised in laws concerning persons in applications of secure access (such as public key), information exchange and data mining. The idea of a VID has already been in use since a long time by doctors who give each patient a pseudo-identity as provision for medical data. [474] Today the digital pseudo-identity is used to assist other doctors, pharmacists and lab assistants in order to facilitate the actions that are required for the health of the anonymous but 'real' patient. With today's advanced technology in hospitals all kinds of autonomous systems will interact.

The personal consequences of having more identities (available) and sometimes using them in an anonymous way seem not to be just fine and fun. On one hand it gives some fun and freedom in exploring your identity and the Internet, but on the other hand the procedures in using electronic authentication for your real identity are becoming more and more stringent. For instance. in South-Korea where game players have to enter their National ID numbers before participating in an online game. [475]

> *'It is better to be hated for what you are than to be loved for what you are not.'*
>
> Andre Gide, French author (1869-1951)

As the virtual worlds' media, including the virtual communities and online game worlds, are also ruled by terrestrial related authorities (moderators, peer group specialists), sovereigns (e.g. administrators and providers) and threats of sanctions (e.g. blocking the access by a specific IP address), often the question will be 'who am I now?' Virtual worlds will become increasingly important in the future, for reasons that reach far beyond games, so the fundamental rights around the individual personality is very important.

So far the three most important personal law topics I can identify are:
1) Privacy and anonymity in virtual environment;
2) Freedom of speech and thought in virtual environment;
3) Ownership of your virtual identity.

---

very amusing when she is likened to dictators simply for kicking people out of "her" channels. Because obviously keeping someone from one channel out of thousands is exactly the same as killing and/or torturing millions of people. WWW: <http://docs.indymedia.org/view/Sysadmin/IrcOper>.

[473] M Bogdanowivz and I Beslay, 'Cyber-security and the future of identity', IPTS report, September, 2002 WWW: <http://www.jrc.es/home/report/english/articles/vol57/ICT4E576.htm>.

[474] The doctor keeps the relationship between the identity and pseudo-identity of the patient secret. He could entrust the identity and corresponding pseudo-identity to a trusted third party. The doctor records the medical data on the patient under his pseudo-identity. Other parties can now have access to the database containing medical information without learning the patient's identity.

[475] 'Korea Produces Safer Online Registration Guidelines' WWW: <http://english.chosun.com/w21data/html/news/200610/200610020023.html>.

The question is not whether or not these rights should be allowed. It is self-evident that they should be. The question is how to deal with the consequences, and what other important rights should be addressed together with these consequences.

The activities of 'intelligent' software agents – with or without virtual identity – will lead to numerous ways of processing personal data, such as the personal data an agent supplies to other agents during transactions, the personal data an agent collects or its user, and the data the agent-provider can extract from the use of his agent. [476] To defend the privacy of the persons implicated, it is important that this personal data is used with caution, that it is necessary for legal purposes, that the data will not be disclosed to the mistaken persons, and that personal data is not processed without the knowledge of the person(s) concerned. Therefore, the use of agents and the processing of personal data have to meet certain conditions that derive from the principles of privacy, which are laid down in most laws and international treaties.

From all these conventions, regulations, and directives, we can abstract the privacy principles, be it:
- Anonymity,
- Purpose specification,
- Legitimate grounds,
- Compatible use,
- Proportionality, and
- Data quality.   These are strongly interconnected.

Designers, developers, suppliers, and/or providers of software agents (with or without virtual identity) must consider these principles while they plan an agent, and must do so in the light of the fundamental right of an individual to decide when and in which circumstances personal data may be revealed.

Do online personalities also have their right of privacy? The rules for data protection are taking away the privacy rights of personas. Your data is processed in a way you don't like. Warren & Brandeis [477] defined in 1890 the right to privacy as 'the right to be left alone'. How to deal with all the security cameras and monitoring systems? Also, more and more the Internet is part of the surveillance, and 'you never can be alone'! In what way the virtual identity source can be used as a tool in preserving anonymity is as yet unknown. But history teaches that the 'nerds' will react in a contrary way and develop their own code.

The EU Ministerial Statement 'where the user can choose to remain anonymous offline, that choice should also be available online' [478], together with the so-called digital cash as trade by barter (as variable of 'give away, take away') enables many Internet users to use at least a real and a virtual identity, even in commercial transactions. 'The various services and activities available over the Internet must be examined, and wherever possible analogies drawn with existing services using older more established modes of communication and means of delivery. Such comparisons will provide a valuable insight into those areas where the possibility to remain anonymous is desirable and those where it is not.' In the second proposal it was reported that anonymity on the Internet would be allowed the same free exposure and manifestation as in the physical world. Europe's John Doe can stay anonymous! [479]

The ownership of a virtual identity will be discussed in the next section.

> *'What I'm doing is building integrated forms of man and machine, and even multiple men and multiple machines, that have as a result that one individual can be super-intelligent, so it's more about making people more intelligent and allowing people to be able to deal with more stuff, more problems, more tasks, more information. Rather than copying ourselves, I'm building machines that can do that.'* [480]

Leaving the handling of your personal contacts and transactions to machines is risky.

---

[476] RN Sobol, 'Intelligent agents and Futures Shock: Regulatory Challenges of the Internet', p 25 (Iowa J. Corp. L. 103 1999).

[477] M Warren & LD Brandeis 'The Right to Privacy. The Implicit made Explicit', *Harvard Law Review* (1890) pp 193-220.

[478] EU Ministerial Conference, Bonn, 6-8 July 1997. 'Anonymity on the Internet'. Recommendation 3/97. WWW: <http//europa.eu.int/ISPO/bonn/Min_declaration/i_finalen.html> and WWW: <http//europa.eu.int/ISPO/eif/InternetPoliciesSite/Crime/PublicHearingPresentations/EuroISPA.html>.

[479] *Wired News*: 'Can John Doe Stay Anonymous'? 21-02-2001. 'What kind of lawsuit do you have when the plaintiff is happy to drop the charges'; WWW: <http://www.wired.com/news/politics/0,1283,41714,00.html> and:

C Kaplan, 'Virginia Court's Decision in Online 'John Doe' Case Hailed by Free-Speech Advocates'. *New York Times* 16-03-2001. WWW: <http://courses.cs.vt.edu/~cs3604/lib/Freedom.of.Speech/Anonymous.html>.

[480] J Brockman, 'Intelligence Augmentation, A Talk With Pattie Maes' (1998); The Third Culture Maes is particularly interested in building autonomous agents that interact with people. Her project allows a user to interact in real-time with 3D animated autonomous characters. WWW: <http://www.edge.org/3rd_culture/maes/index.html />.

The risks are divided into two groups: risks caused by agents acting on behalf of a user and risks caused by foreign agents that act on behalf of others. The potential and implications of using them are hardly well-considered. By using the 'privacy enhancing technology' an overall solution for a privacy 'respecting' agent is given. [481] Using the software agents as representative in transactions is shown in figure 37. Another step is to extend the individual citizen's view to a political view asking for 'democracy-enhancing technologies', think of freedom of speech, or of information.

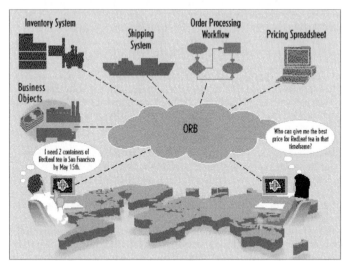

The legal aspects of VIDs facilitated by intelligent agents and multi-agent systems concern the legal nature of intelligent agents and multi-agent systems, civil and criminal liability of agents in online games, e-commerce, knowledge management and business networks, use of information collected by agents, broker arrangements by agents, negotiations and e-contracts made through (software) agents, et cetera, either in a single stand-alone situation up to in an advanced multi-agent environment.

Figure 37: eCommerce using various Intelligent Agents and an Object Request Broker (ORB) [482]

Humanitarian Law

One of the main principles of International Humanitarian Law, target discrimination, is used to highlight the legal and ethical dilemmas involving the use of autonomous agents in relation to human beings.

Zwanenburg, Boddens Hosang, and Wijngaards [483] discuss the humanitarian dilemmas in target discrimination when using autonomous intelligent agents in war equipment such as guided missiles.[484]

> 'On July 3, 1988, the USS Vincennes, a Ticonderoga Class cruiser equipped with the highly sophisticated Aegis system, was fighting and engaging multiple targets at once, a contact thought to be an Iranian F-14 aircraft was seen to be inbound to the Vincennes. Assumed to be on a descending course straight for the ship, the air contact was tagged hostile and ultimately engaged and destroyed. The contact later turned out to be a civilian Airbus type aircraft, Iran Air flight 655 from Bandar Abbas to Dubai, carrying roughly 290 passengers. There were no survivors. ... The system was designed to provide extensive battle space management and enable area defence against multiple air, surface and sub-surface targets. The level of autonomy with which the system could engage targets could be increased or decreased according to the threat level and circumstances.'

---

[481] R Hes and J Borking (eds), 'Privacy enhancing Technologies: The path to anonymity'. Rev. Ed. A&V-11 (1998). Den Haag: Registratiekamer. WWW: <http://www.ipc.on.ca/images/Resources/anoni-v2.pdf>.

[482] P Fingar, 'Intelligent Agents: The Key to Open eCommerce; Building the next-generation enterprise. WWW: <http://home1.gte.net/pfingar/csAPR99.html>.

[483] M Zwanenburg, H Boddens Hosang, and N Wijngaards, 'Humans, Agents and International Humanitarian Law: Dilemmas in Target Discrimination'. In: Proceedings of The 4th Workshop on the Law and Electronic Agents (June 2005) pp. 45-51. WWW: <http//www.lea-online.net/publications/ZwanenburgBoddensWijngaards_LEA05_CR.pdf>.

[484] The principle of discrimination basically states that combatants and military objectives may legitimately be attacked, and that it is prohibited to attack non combatants and civilian objects. Its codification in Article 48 of Additional Protocol I ('In order to ensure respect for and protection of the civilian population and -objects, the Parties to the conflict shall at all times distinguish between the civilian population and combatants and between civilian objects and military objectives and accordingly shall direct their operations only against military objectives.') is complemented by a number of other provisions in the Protocol which relate to a number of consequences of the distinction, including Article 51(2) which prohibits attack of civilians and Article 52 which defines military objectives. Of central importance to the application of the principle of discrimination to agent technology is the interpretation of the words 'effective contribution to military action' and 'definite military advantage' in Article 52. 'Those objects which by their nature, location, purpose or use make an effective contribution to military action and whose total or partial destruction, capture or neutralization, in the circumstances ruling at the time, offers a definite military advantage.'

In a tense situation in a combat zone, an unfortunate mistake caused the deaths of several scores or hundreds of people. It has happened in every war, and will happen again in the next. There appears to have been little interest in it as a case of malfunctioning technology. [485]
The virtual identity in law of persons is developing steadily, almost parallel to the development of software agents with artificial intelligence and autonomous systems. It can be concluded that there is no legal but only a medial difference between an identity in the real world and one in the cyberworld. However, it will take time and building brick-by-brick jurisprudence to bridge both identity-worlds. Support on unimpeachable authority will help us: the cross-media technology leads the way to integrating both physical (journals, tools) and virtual (Internet, games) environments.

VID in Law of Property

There are some relations between VIDs and property, mainly in property rights.
In some – mostly money driven – cases (Reynolds) [486] the owner wants to claim the intellectual property of his avatars. In other cases the user of a VID in the virtual environment never wants to be associated in real life with 'that' virtual identity. This makes clear the popularity of anonymous participation in blogs, chat rooms and virtual worlds. In some situations that cannot acceptable, the VID executes harmful actions that are not welcome by the victim.
In many cases the law has established that the copyright in software subsists in code. But the way an avatar is displayed (on a screen) interacts with other system elements that are controlled by code; what individuates a specific avatar is not this code (which is common) but a set of data entries stored in a database or otherwise (Lee) [487]. Hence rights in the game software do not apply to any given virtual appearance. Also, when a player creates and uses a virtual identity such as an avatar (s)he is not creating a piece software subject to copyright either.
Rights of publicity try to capture the relationship between identity and expressions of persona that resonate with the relationship between the avatar and the individual. These IP rights are outdated and wholly alien to what today happens in the virtual world and in the cross media environment. Law follows society, but in this particular situation the law did not 'log in' at all. As Jacoby and Zimmerman put it: does it make sense that Tiger Woods could (in theory) own the persona of Michael Jordan? [488]

The economic aspect of virtual items such as avatars and identities also has tended to frame the debate about them in terms of property and an intersection between items, disputed acts, code and the law as it stands (Lastowka and Hunter)[489]. The law most commonly associated with property disputes over virtual world items is the copyright law. Reynolds suggests that 'developer-publishers believe that they have a natural property right in virtual items as they create virtual worlds so own every aspect of them. Developer-publishers also tend to believe that the control that property rights grant them is needed for bringing coherence in a virtual world because that is necessary for the good of all players. Many players also believe that they have a natural right of property in both virtual items and especially avatars and identities, this stems also from law and the view that as there is no avatar in the box when the game is purchased so avatars must be created through the application of player effort, hence from labour-desert theory this is naturally their property.'

Keeping the ownership of your own virtual identity is a serious concern. Perhaps some things as virtual identities should not be understood as property. Comparing with the law in real life, the cases often referenced are *Midler v. Ford* and *Waits v. Frito Lay* [490] which have been interpreted as adding a *'widely known'* test to publicity rights claims such as in which is held that 'widely known' could be defined as *'known to a large number of people throughout a relatively large geographic area'*. If a *'widely known'* avatar does not look as an individual it could never fall under these rights. However some courts have ruled that likeness is not a requirement. The key case here is *Motschenbacher v Reynolds* [491] which examined –under Californian state-law – whether the use of

---

[485] GI Rochlin, 'Trapped in the Net: The Unanticipated Consequences of Computerization' -- Chapter 9: 'Unfriendly Fire'. WWW: <http://press.princeton.edu/books/rochlin/chapter_09.html>.

[486] R Reynolds, 'Hands off MY avatar ! Issues with claims of virtual property and identity' WWW: <http//www.ren-reynolds.com/downloads/HandsOffMYavatar.htm>.

[487] J Lee, 'Data-Driven Subsystems for MMP Designers: A Systematic Approach', in: *Game Developer* (2003) V.10/ 8, pp 34-39.

[488] MB Jacoby and DL Zimmerman, 'Foreclosing on Fame: Exploring the uncharted boundaries of the right of publicity', *New York University Law Review*, Vol 77 No5. (2002).

[489] GF Lastowka and D Hunter, 'To Kill an Avatar', *Legal Affairs* (2003); WWW: <http//www.legalaffairs.org/issues/July-August-2003/feature_hunter_julaug03.msp>.

[490] Midler v. Ford, 849 F. 2d 460 (9th Cir. 1988). Waits v. Frito Lay, 978 F. 2d 1093 (9th Cir. 1992).

[491] Motschenbacher v. R.J. Reynolds Tobacco Company, 498 F.2d 821 (9th Cir. 1974).

a distinctively marked car in a commercial infringed upon a driver's right of publicity. Finding that it did, the judge stated 'that the driver is not identifiable as plaintiff is erroneous' and – regarding the relationship between the markings and persona – 'these markings were not only peculiar to the plaintiff's cars but they caused some persons to think the car in question was plaintiff's and to infer that the person driving the car was the plaintiff. Therefore it is conceivable that an avatar would be ruled in scope as people that interact with an avatar do identify it with an individual. 'One who appropriates the commercial value of a person's identity by using without consent the person's name, likeness, or other indicia of identity for purposes of trade is subject to liability.' [492]

Bygrave [493] deals with a second important concern: the relation between Digital Rights Management (DRM) and privacy ([493] pp 418-446). In his opinion 'recent developments in Digital Rights Management Systems (DRMS) are bringing to the fore considerable tension between the enforcement of intellectual property rights and the maintenance of consumer privacy' ([493] p 418). Hence what is required seems to be an integration of technological measures for protecting intellectual property rights with privacy enhancing technologies (PETs). More precisely Bygrave recommends building mechanisms into DRMs architecture which enhance the transparency of the systems for information consumers, and building mechanisms into the systems architecture which preserves, where possible, consumer anonymity, and which allow for pseudo-anonymity as a fall-back option, i.e. a separate persistent virtual identity, which cannot be linked to a physical person or organisation. In parallel, as he says, 'it may be useful to draw on the technological-organisational structures of DRMS to develop equivalent systems for privacy management' ([493] p 446). In short, the development and application of the 'least privacy-invasive devices' is encouraged to preserve the value of a virtual identity.

However, Carpenter [494] argues that' the 'identifiably' test has been so inconsistently applied that is has become muddled and completely ineffective'. Nevertheless, in this context identity can be seen – as legal security; – as private life; and/or – as property. He argues too that the phenomenon of amateur Internet publishing, with its low cost, accessibility, and broad audience, calls for a new and expanded concept of 'commercial value' as it is understood within the context of the right of publicity.

Unfortunately, companies are now faced with employees and former employees making disparaging comments about them online [495]. Such criticism and negative comments are referred to as 'cybersmearing'. The number of such lawsuits is dramatically on the rise.

> Dec. 14, 1999 — It was the end of another long day at her home office when Amy, a New York City consultant, discovered to her horror that she had been enrolled in a small but growing club that no one joins voluntarily - victims of a 'cybersmear.' 'I did not sleep the entire night, mostly because there was nothing I could do,' she said of the hours after she discovered a Web page bearing a doctored photo that falsely showed her engaging in sexual relations. '... I felt ... dirty and abused and violated.'
>
> Brunker, 'Cybersmeared' -- One Victim's Tale' MSNBC 1999

SecondLife, one of the popular Massive Multiplayer Online virtual worlds, have recently met the phenomenon of a (virtual) terror attack by (virtual) terrorists. [496]

> The Second Life Liberation Army (SLLA) is detonating computer versions of atomic bombs in a campaign to establish universal suffrage in the world, which is inhabited by more than four million people from across the world. According to its Website, the SLLA is the 'military wing of a national liberation movement within Second Life' dedicated to establish democracy, handing residents a vote in the direction of the world. It accuses Linden Labs as 'functioning as an authoritarian government' and the only 'appropriate response is to fight'.
>
> According to the Australian newspaper The Inquirer a terror campaign has been waged in Second Life which has left a trail of virtual dead and injured, and caused hundreds of thousands of dollars' damage. Apparently there are three jihadi terrorists registered and two elite jihadist terrorist groups in Second Life and they use the site for recruiting and training.

---

[492] Restatement of the Law (Third) of Unfair Competition, USA (1995).

[493] L A Bygrave, 'Digital Rights Management and Privacy - Legal Aspects in the European Union'. In: E Becker et al. (eds) Digital Rights Management, LNCS 2770 (2003) pp. 418–446. WWW: <http://folk.uio.no/lee/publications/DRM_privacy.pdf>.

[494] JL Carpenter, 'The Case for an Expanded Right of Publicity for Non-Celebrities; Virginia Journal of Law and Technology' V. 6/3 (2001); WWW: <http://www.vjolt.net/vol6/issue1/v6i1a03-Carpenter.html>.

[495] MD Risk, JS Barber, SR Gallagher, JA Totten, and SE Fox, 'Cybersmear It's What The Internet Is For, Right?' (2005); WWW: <http://www.bna.com/bnabooks/ababna/tech/2005/barber.pdf>.

[496] 'Terrorists attack Second Life' (26-02-2007) <http://www.nowpublic.com/node/213622>. Secondlife, WoW plagued by terrorists (01-08-2007); WWW: <http://theinquirer.net/?article=41361>.

In April of 2007 Lindenlab, the US based company and owner of SecondLife, contacted the FBI to turn over the names of the miscreants who had crashed their Massive Multi Online (MMO) Website. Shortly before, they had asked US officials advice on the position of the online Casinos in their virtual world. This action led to violent blog discussions in many countries, e.g.:

> 'Some people shut down the MMO Second Life by creating self-replicating objects with the in-game scripting system that replicated so much that they crashed the server. The CEO of Linden Labs (the game's publisher) turned over the names the grievers to the FBI. I'm usually quick to point out the problems of banning people who perform legal in-game actions that have consequences that 'you don't like.' In almost all of these cases, good answers are 'have the developer fix the problem in the code' or possibly 'allow players a way of policing or otherwise sort out the problem themselves.' This is an extreme case though, since these actions caused the entire server to crash, denying all other players the ability to play and the company the ability to make money. Perhaps in this extreme case, it *IS* correct for the developer to step in and ban, and call the authorities. Almost anything less than this probably does not warrant a ban or any penalty to the player from the developer.' [497]

Basically, the most current virtual worlds such as World of Warcraft have medieval hierarchy, where players have nearly no rights at all. User generated content, code transparency, and VID-democracy will change the virtual world. In a hundred years, this type of online government will be a cute footnote.

The Position of the Virtual in the Real World System of Rules

In order to call up the virtual effects (by software) that created an imaginary world for us that flitted through our brain there are rules to obey from at least four sources:

1. *Private Agreements*. The first source occurs in deals between the people involved. These rules suggest what each person will do and will not do. These implications are agreements or promises. The agreements and commitments here can be one-way or two-way. They may be written or spoken. People get upset whenever such rules are not obeyed or not understood. To prevent that upset feeling we should adapt our behaviour.

2. *Public Laws in Real Society*. A second source is given by rules or laws that express what is acceptable in the society. These govern in part our behaviour. Such laws say what we should or should not do. They may even specify penalties or punishments for disobeying them. A rule that is not enforced, or is enforced weakly, is often ignored or forgotten.

3. *Physical Laws*. The third source of rule occurs in technology, mathematics and science. These record or state our observations of nature and the patterns it follows. They may describe what has been seen. They record human experience. Examples of the latter are provided by the recipes for cooking and operating instructions for machines. Reliable and carefully followed procedures give reproducible results. Furthermore, recipes and reliable patterns can be joined together to suggest more recipes and patterns of behaviour.

4. *Application Laws*. A fourth source is found in the source code of the software. The application controls the use of that application by offering options in driblets. Using this way of regulated feedback results in an adapted desirable behaviour, as welcomed behaviour is rewarded and unwelcome behaviour is restrained.

These four sources of rules or patterns are discussed later on in parts.

Interaction and Reciprocity

Private arrangements belong to the merchant community with a market system of property rights and rules of exchange. The commercial sector is completely capable of establishing and enforcing its own laws. Commerce is an evolving system of commercial law. Menger [498] proposed that the origin, formulating and the ultimate process of all social institutions including law is essentially the same as the 'spontaneous order' Adam Smith described for markets.[499] Markets guided by Smith's invisible hand coordinate interactions, and so does customary law. These systems develop because through a process of trial and error it is found that the actions they are intended to coordinate are performed more effectively under one process than under another one. The more effective institutions and practices replace the less effective ones.

---

[497] 'Second Life calls the FBI'; WWW: <http://www.sirlin.net/archive/second-life-calls-the-fbi/>.

[498] K Menger, *Principles of Economics* (1871). Translated by J Dingwall and BF Hoselitz, with an introduction by FA Hayek (New York University Press,New York 1981).

[499] A Smith, 'The Theory of the Moral Sentiments' (1759). WWW: <http://libertariannation.org/b/spont.htm>.

Customary Law in Cyberspace

After implementing the mediation programme they instituted for eBay, Katsh et al [500] (p 728) report discovering that: 'As we encountered disputants and observed them as they participated in our process, we began to see eBay not from eBay's perspective, which assumes that eBay is the equivalent of a landlord with little power over how a transaction is finalised, but from the user's perspective. The more we saw of this, the more we became persuaded that disputants were, indeed, participating as if they were *'in the shadow of the law.'* The law whose shadow was affecting them, however, was eBay's law rather than the shadow of any other law.' [501] Thus, eBay is not just a marketing arrangement, but it also is a legal jurisdiction. Parties agreed to participate in mediation 'at a very high rate' because of eBay law. Their primary concern was in maintaining their eBay reputations. As Katsh et al ([500] p 729) explain, Ebay's response to this public safety problem was not to install a police force to deal with problems after they occurred but to use an information process to try to prevent disputes from occurring at all. Since the public safety problem largely focused on unknown and perhaps untrustworthy sellers and buyers, eBay put in place a process for sellers and buyers so that they could acquire reputations as trustworthy parties. Protecting one's feedback rating looms large in any eBay user's mind. As one guidebook to eBay points out, 'on eBay, all you have is your reputation'. Katsh et al ([500] p 731) argue 'While online auctions try to limit potential liability by creating distance between the auction site and those doing business in the auction site, the site owners are the designers and administrators of the process of creating identities and establishing reputations. This is a formidable power and, while it might appear that the auction site owners are merely making a process available and then letting users employ it, there are terms and conditions governing these data collection and data distribution processes, and these rules are made and administered by eBay and other proprietors of auction sites'.

Economy and Law

The basis of intellectual property (IP) law is not only a-territorial, but is also related – private law in particular – to the economy. Salzberger [502] states that three variables construct the basis of law and economic analysis: *communities, time* and *individuals*:
He points out to Demsetz [503] when he argues that in Demsetz' original analysis, which focused on natural resources like hunting land, oil or waters, it was sensible to define the community on the basis of territory. This is not the case with IP and a Public Domain of ideas. Likewise, the implicit assumption of the incentives model that the unit of maximisation ought to be the state (as it advocates IP laws enacted by the state) is far from being self-explanatory. Ideas cross territorial and political boundaries. IP markets are global. Intellectual community activities are a-territorial. The implication of the borderless nature of ideas on economic analysis is highly significant. One can no longer take the state as the relevant framework for market activities, for decision-making calculus or for institutional analysis. This change is significant in both the normative and positive domains. The second variable that ought to be defined in order to conduct maximisation of welfare, wealth or utility is a *time* framework. The definition of time is less acute when economic models analyse responsibility rules for physical harms or criminal law. It is very significant when dealing with a propriety regime and especially when we analyse intellectual property. In cyberspace, time is the message. (Kokswijk) [504]

Salzberger's third important variable, which constructs the basis of law and economic analysis, is the individual. Citation: 'Most models assume that individuals are rational physical entities and each has a fixed set of preferences or a utility function, which is exogenous to the object analyzed by the model. In other words, these preferences are pre-fixed and do not change as the result of

---

[500] E Katch, J Rifkin, and A Gaitenby, 'E-Commerce, E-Disputes, and E-Dispute Resolution: In the Shadow of 'eBay Law', Ohio State Journal on Dispute Resolution, 15(2000) pp 705-734. WWW: <http://www.umass.edu/cyber/katsh.pdf>.

[501] The terminology, 'in the shadow of the law' is generally attributed to Mnookin and Kornhauser (1979: p 968), in: R Mnookin and L Kornhauser 'Bargaining in the Shadow of the Law: The Case of Divorce', *Yale Law Journal*, p 880; who suggested that bargaining occurs in the shadow of the law because the legal rules give each party 'certain claims based on what each would get if the case went to trial. In other words, the outcome that the law will impose if no agreement is reached...' Since then it has often been contended that ADR also other sources of law might cast some operates in the shadow of the law, where law implies state-made law (statutes, precedent).
As Katsh et al (2000: p 728) imply, non-state made law (i.e., customary law) also casts a shadow. They acknowledge that shadows too, but they are not very significant if they do. Recourse to state-made law and public courts is rarely even mentioned.

[502] EM Salzberger, *Economic Analysis of the Public Domain*; in: 'The Future of the Public Domain', Ch. III, pp. 27-59, (Kluwer Law International, 2006).

[503] H Demsetz, 'Towards a Theory of Property Rights' 57 *American Economic Review* (1967) 347-360, in 4.3 'The arena'.

[504] 'In cyberworld Time is the Message'. In: J van Kokswijk ,Architectuur van een Cybercultuur (Bergboek, Zwolle 2003) and in: J van Kokswijk, Hum@n, Telecoms & Internet as Interface to Interreality (Bergboek, 2003; http://www.kokswijk.nl/hum@n.pdf>).

deliberation and interactions within and outside the relevant market. Two major points can be highlighted in context of this fundamental presupposition. The first relates to the definition of the individual in the new information environment; the second is connected to the debate between liberal and republican theories of the state.

The new Information environment transforms not only the notion of collective communities, but also that of the individual, who is the basic unit for liberal philosophy of the state and for economic analysis. In the non-virtual world the basic unit of reference – the individual – is one person with a single identity, passport or drivers license number, a specific address and distinct physical features. In the new information environment, the atomistic unit of analysis is a username with a password and an electronic address. There is no strict correlation between the cyberian individual and non-virtual individual, as the same physical individual can appear on the Internet as several entities, each with different identification features and a different character, belonging to different communities.

While conventional economic thinking, perceives individual preferences in the non-virtual world as exogenous to the political process and to the economic markets, the new information environment requires us to internalise even the analysis of individual preferences. Conventional economic analysis assumes that our basic identity, which can be framed in terms of various sets of preferences, is the result of distinguished historical, cultural, linguistic and even climatically different backgrounds (Montesquieu) [505]. 'Those background factors are pre-given and pre-date any formation of markets and collective action organisations, such as states or other national units. The definitions of state boundaries, however, are very much influenced by these ancient groupings of preferences. Even if preferences change as the result of market interactions, such as successful marketing and advertising, they are initially founded upon these ancient differences, some of which are presumably almost permanent.'

With technology making rapid strides lawmakers have to frequently change their application systems due to technology obsolescence. This challenge to the law is even more prominent in the 21st century, wherein the relationship between law and technology has assumed a sharper focus. Now, an age-old recurring controversy in the history of legal thought has been resurrected, controversy between those who believe that law should follow the social sentiment, norms and those who believe that law should be the determining agent in the creation of new social norms.

Rifkin argues in his 'Age of Access' that social ordering will be more influenced by access to services, experience and networks rather than property relations. [506] Technology has always posed problems for the law but lawyers and judges have been managing the problems by stretching the meaning of the existing laws without breaking the spirit of laws. Since they had time on their side and technologies progressed at a leisurely pace, managing problems posed by technology was not difficult. But today with the coming of digital technology, unusual problems are being presented so much so that it has set into motion a historical force, which has come to be known as the 'Digital Revolution', a 'convergence of computer networks and communication technologies'. This digital revolution has given birth to cyber-space, which itself has ushered in the knowledge economy wherein, in addition to capital and labour, knowledge has become the third factor of production. Since the knowledge economy revolves around information, the nature of information raises very important issues which the law has to contend with. Information is intangible, *res nullius*, and is the common heritage of mankind. Efforts to cabin it and confine it through proprietary forms of ownership must be justified in the larger public good.

Hidden Habitants in the Heat of Hyperlinks

This subchapter points out the topic of feeling intangible when being anonymous in cyberspace. It overviews the ways criminals use the virtual identity as cover for illegal activities. It is observed in what way organised crime is in the lead.

The Attraction of Anonymity

---

[505] C Montesquieu, *The Spirit of Laws*. [1748] (University of California Press, Berkeley 1977).

[506] J Rifkin, *The Age of Access: The New Culture of Hypercapitalism, Where all of Life is a Paid-For Experience* (Tarcher, 2001).

Can an Anonymous Virtual Identity Really be Anonymous?

This hypothetic question has two sides, technological and legal. The answer is a dual negative: Both technology and legal investigation will link the VID to something, a desktop PC or so. Linking to someone, a human being, is not self-evident.
With today's technology almost all connections via the Internet can be monitored and the IP addresses connected to the network device of a computer (notebook, desktop, server, no matter what) are logged. Even with use of a proxy or anomyser, there will be always traces that lead to the originated computer.

> '... No court could fail to notice the extent to which business today depends on computers for a myriad of functions. Perhaps the greatest utility of a computer ... is its ability to store large quantities of information which may be quickly retrieved on a selective basis. Assuming that properly functioning computer equipment is used, once the reliability and trustworthiness of the information put into the computer has been established, the computer printouts should be received as evidence of the transactions covered by the input.'   Harris v Smith, 372 F.2d 806 (8th Cir.1967)

Knowing the location of the network-terminal (such as the physical address and floor/room) and the characteristics of a connected modem (such as the network device MAC-Address) will present that someone did 'what, where & when' but do not solve the evidence of 'who' did it. [507] More traces can be found in the used computer itself, for example in log files and in the memory.
Sometimes it is necessary to use legal procedures in order to disclose the customer's name and address. In most cases such a claim is sustained, but sometimes it is not.

> Supreme Court of Virginia refused to grant a unidentified company access to America Online's confidential subscriber information unless the firm agreed to reveal its identity. [508] Rural/Metro, an ambulance and fire service company in Scottsdale, Arizona, US, sued four individuals who had posted messages on the company's Yahoo finance message board that contained what it alleged to be confidential and libellous material. Privacy and free speech advocates saw the Rural/Metro case as a missed opportunity. Had the case gone to trial, the court would have addressed whether, or under what circumstances, Internet Service Providers would be required to divulge private information. In the suit, Rural/Metro subpoenaed Yahoo, demanding the identities of the four individuals involved in the postings.

However, the physical location and the identity of the owner of the connection and/or computer do not prove the identification of the initiator and/or user of a VID. In many situations digital traces compared with patterns of human behaviour can give more evidence of one's identity.

In the physical world, more and more countries regulate in conventions and treaties the cross-border surveillance and investigation of police.[509] As the Internet has no visible borders, enforcement authorities from many nations cooperate 'virtually' in the fight against cybercrime by (1) swapping information and (2) cooperating in the seizure of evidence on local computers. [510] The cooperation can be by treaty, by informal cooperation and by cross-border searches and seizures. [511] Goldsmith states that even in the absence of treaties, a nation can act unilaterally: Sitting at their desks in one country, law enforcement officials can take unilateral steps on computer networks to trace the origins of the cyber attack and explore, freeze, and store relevant data located on a computer in the country where the crime originated. [512]
These actions are known as remote cross-border searches and seizures.

---

[507] In computer networking a Media Access Control address (MAC address) is a unique identifier attached to most network adapters (NICs). It is an identification that acts like a name for a particular network adapter, so, for example, the network cards (or built-in network adapters) in two different computers will have different names, or MAC addresses, as would an Ethernet adapter and a wireless adapter in the same computer, and as would multiple network cards in a router. However, it is possible to change the MAC address on most of today's hardware. Thus, the unique identification disappears.

[508] 'Virginia Court's Decision in Online 'John Doe' Case Hailed by Free-Speech Advocates', New York Times, 16-03-2001; WWW: <http://courses.cs.vt.edu/~cs3604/lib/Freedom.of.Speech/Anonymous.html>.

[509] Institute for International Research on Criminal Policy; 'Study on cross-border and international cooperation between police services'; WWW: <http://www.ircp.org/uk/category.asp?category=cat6>.
In 1985, the Netherlands, Belgium, Luxembourg, Germany and France decided to set up permanent cross-border crime teams with local police in several towns. WWW: <http://www.mindef.nl/binaries/kmar_tcm15-23067.pdf>.
Border crossing policemen keep their qualification as criminal investigator. Treaty between the Kingdom of Belgium, the Kingdom of the Netherlands and the Grand Duchy of Luxemburg concerning cross-border police intervention. 08-06-2004. WWW: <http://www.benelux.be/en/pdf/rgm/rgm_Politieverdrag2004_en.pdf>.

[510] Computer Emergency Response Team, abbreviated to CERT or GOVCERT, to protect the nation's Internet infrastructure.

[511] Historical and anthropological evidence suggests that the earliest men lived in groups that were largely cooperative. Source: Robert Ellickson, 'Property in Land', Yale Law Journal, vol. 102 (1993): pp 1315-1400.

[512] JL Goldsmith, 'The Internet and the Legitimacy of Remote Cross-Border Searches' (2001): Forthcoming Available at SSRN. WWW: <http://www7.nationalacademies.org/cstb/wp_cip_goldsmith.pdf>.

Goldsmith points out that there are two problems with these unilateral acts. First, many believe they violate the principle of territorial sovereignty and thus violate international law. Referring to the legitimacy of remote cross-border searches over the Internet Goldsmith argues, contrary to conventional wisdom, that cross-border searches and seizures are consistent with international law. And second, cross-border searches cannot produce the criminal defendant himself.

The Legally Intangible Anonymous Identity in the Cyberworld.

Computer software designers created models of identity systems where every real life entity can have multiple virtual identities. Dewan and Dasgupta state that in well designed models there is no way of tracing the real identity of an entity from its virtual identities. [513] In addition, virtual identities are independent of each other, i.e., neither the action performed under any of the virtual identities nor the real identity does not impact the other identities of the subject.

> **An example** of such a system is the Yahoo chat room where one can log-in with multiple identities (with multiple clients), and all the actions performed by each virtual identity are totally independent of the other virtual identity or the real life identity. There is no method by which any of Yahoo identities can be traced back to the real life subject. Only a session of a virtual identity can be traced back to the IP address, from where the entity is logged on the chat room. As already mentioned, the gap between the machine and the human is too big to fill. Hence the probability of tracing the real entity is nil, given the fact that the entity does not disclose any other information than what is required by the chat room. Yahoo requests the real life information of the entity, but does not validate it.

It could be that the gap between the machine and the human 'behind the controls of that machine' is too big to fill, and it could be that there are no footsteps traceable such as in real world, but – when the investigator has found a clue to an IP address – there are always tracks and patterns that leads to specific human profiles. One of the most identifiable tracks is the use of codes and files in a connection, such as the electronic fingerprints in digital pictures and the hidden registration codes in used software. Even the characteristics of programming a DOS or designing a bot are – much like handwriting – deducible to a person. Also, identifiable are the characteristics in behaviour and in the use of words and images in a conversation.

Human behaviour is often predictable and someone's conversation is identifiable, such as the regular intervals, focus of the edits, kind of debate, comments to users, familiarity with an obscure subject, in-depth knowledge, the style of writing, and/or the slightly awkward, use of a foreign language. Also, the way and the frequency are indicators.
After observing email and BGP traces Duan et al conclude that spammers are more likely to rely upon the technique of using network prefixes with a short life duration rather than on network prefixes with persistent short reachable intervals. [514]
In 1990 the 'Guideline for Federal Records Managers or Custodians' was prepared, [515]

> 'These patterns of behaviour provided us with a general description of the intruder -- we knew his modus operandi, his hangouts, his patterns of computer speech, the computer tools he used for his break-ins, and his disguises,'

> 'We intercepted only those communications which fit the pattern,' explained Stern. 'Even when communications contained the identifying pattern of the intruder, we limited our initial examination to 80 characters around the tell-tale sign to further protect the privacy of innocent communications.'

> 'We will work with our foreign counterparts to achieve justice,' ... 'International teamwork is being applied to international crimes,'   (cited: the Attorney General)[516]

followed in 2001 by the manual 'Searching and Seizing Computers and Obtaining Electronic Evidence in Criminal Investigations'. [517]

---

[513] P Dewan and P Dasgupta, 'Trustworthy Identification in a Privacy Driven Virtual World, Security Topic Area', *Distributed Systems Online*, January 2004, pp 1 - 8.

[514] Z Duan, K Kopalan, and X Yuan, 'Behavioral Characteristics of Spammers and Their Network Reachability Properties' (2006); WWW: <http://www.cs.fsu.edu/research/reports/TR-060602.pdf>.

[515] 'A Guideline for Federal Records Managers or Custodians' (1990). U.S. Dept of Justice, Justice Management Division, Systems Policy Staff, in: 'Admissibility Of Electronically Filed Federal Records As Evidence'. WWW: <http://www.lectlaw.com/files/crf03.htm>.

[516] Federal Cybersleuthers armed with first ever computer wirer tap etc. Argentine man accused of breaking into Harvard University's computers (1996); WWW: <http://www.fas.org/irp/news/1996/146.htm>.

[517] Computer Crime and Intellectual Property Section, Criminal Division, July 2002; United States Department of Justice. WWW: <http//www.cybercrime.gov/s&smanual2002.htm>.

Certainly, bare statements such as 'There's no way that the computer changes numbers' are likely to satisfy most courts making evidentiary decisions. Notwithstanding, several situations have proven that electronic data can be altered intentionally or unintentionally - with comparative ease. (Vinhee)[518] Unlike paper records electronic information is stored in a networked matrix of computer hardware and software that constantly changes as the software is patched and updated and the hardware is upgraded, replaced and shared. It is important to rule and to demonstrate security procedures, as well as security and data validation features in corporate computer systems when sponsoring electronic documents.

Traces of Terminal Use
Kimmelmann explores how individuals are personally extended through their virtual activities.[519] Just as handwriting is inherently an extension of the person, the portrait inevitably a meditation in the form of a painter's skill, he contributes to the possibilities of authenticity in the electronic media. But is that satisfying the courts?

Just as former U.S. Supreme Court Justice Potter Stewart once said that he could not define pornography, as he said that '*I know pornography when I see it*', it is possible that the same could be said about defining the relation between the anonymous virtual identity and the (real) identity of the real person that controls the machine that was connected by IP to the system in which the anonymous virtual identity exist.
An analogy can be found in the formula '*I know art when I see it.*' and '*I know spam when I see it.*'. It is hard to create a set of rules that would be accepted by everyone to define spam, but each individual may know what he or she considers being spam or cybercrime when he or she sees it. That means that the best way to categorise spam or cyber crime is for each person to categorise it by example. Same, even with an absent consensus definition, most people 'know terrorism when they see it'. Similarly can be argued that due to the behavioural patterns of an identity the court judges '*I know the identity when I see it*'.

The conclusion is that an anonymous identity in the cyberworld is – sooner or later – always legally tangible. The development of new investigation methods and tactics should be more multidisciplinary oriented and be more open to behavioural lead.

Use of Virtual Identity as a Cover for Criminal Conduct

Follow the Virtual Money to the Source of the Crime
As soon as one government cracks down on some unwanted Internet activity within their boundaries – such as voluntary transactions like gambling and pornography, or involuntary externality-created activities like spam, viruses, fraud, and identity theft – the perpetrators can simply set up their operations in other geographic locations. Internet gambling sites are located in liberal, exotic or corrupt states and the owners of these gambling establishments can still sell their services to millions of consumer in other countries where gambling is limited. Spammers have chosen similar routes to avoid detection or prosecution. In fact, some countries, like China, actually encourage externality generating cyber activities such as spam because of the local economic benefits that arise. (Zeller p 2)[520]
Some corrupt governments even encourage other kinds of cyber crime 'as a developing industry' (Rustad)[521]. As Zeller (p 4) points out, for instance, 'The overlapping and truly global networks of spam-friendly merchants, e-mail list resellers, virus-writers and bulk e-mail services have made identifying targets for prosecution a daunting process. Merchants whose links actually appear in

---

[518] In re: Vinhnee, 2005 WL 3609376 (B.A.P. 9th Cir. Dec. 16, 2005, note 9) stands (among other things) for the proposition that a sponsoring party cannot rely upon judicial notice to fill gaps in the explicit foundation that it uses to authenticate electronic materials it wishes to introduce as evidence. It rejects the approach ordinarily used to authenticate computer records.   >>>
DF Axelrod, 'Are More Stringent Rules for Authenticating Electronic Records Coming?' (2006); WWW: <http://www.law.com/jsp/ihc/PubArticleIHC.jsp?id=1150720529549> and:
S Mason, 'Authentication of electronic evidence' (2006); WWW: <http//www.infoage.idg.com.au/index.php/id;1288698068;fp;4;fpid;675408222>.

[519] B Kimmelmann, 'Retexting Experience: The Internet, Materiality, and the Self', in: 'Virtual Identities: The Construction of Selves in Cyberspace'; by C Maun and L Corrunker (L. Eastern Washington University Press 2007).

[520] T Zeller Jr, 'Law Barring Junk E-Mail Allows a Flood Instead' *New York Times*, nytimes.com February, 2005 p2, p4.

[521] ML Rustad, 'Private Enforcement of Cybercrime on the Electronic Frontier,' *Southern California Interdisciplinary Law Journal*, 11 (2001) p 86 in pp 63-116.

junk email are often dozens of steps and numerous deals removed from the spammers … and proving culpability is just insanely difficult.' Anonymity is further enhanced because spammers use viruses to gain control of PCs to use as 'zombie spam transmitters' so even if spam is traced to a particular computer, its owner may be unaware of the activity and completely innocent of any wrongdoing. Governments are not only attempting to control spam, viruses and worms, but also to prevent fraud and identity theft [522]. Their impacts in these areas are similar to their 'success' against gambling and pornography, as many of the viruses, worms, identity theft and fraud activities that attack individuals all over the world emanate from countries where governments are unwilling or unable to stop these activities.

The development of the useful technology of Artificial Intelligence (AI) leads to a huge emergence of AI-bots and 'intelligent agents' [523] in both the criminal and anti-criminal activities (Mena)[524]. These software agents can not only steal an identity but also replace it with a virtual identity, keeping the perpetrator anonymous.

> 'We have known two centuries of endless conflict. Two centuries in which the Commonwealth has failed to bring justice to a troubled universe. The situation is critical and duty must prevail: your duty. This is the time to focus the power of the mighty Dreadnaught and make a difference in humanity's infinite war. This is the time when your destiny becomes your own.'   I-War: Enter Infinity [525]

The real 'i-war' phase, the use of advanced intelligent agents in the information warfare, has already been reached (Busuttil and Warren; Busutelli) [526]. Computer nerds, criminals and crime fighters are using the same technology.

Much of the 'law' is concerned with security, or 'public safety', of course, but in cyberspace the issue is not one of physical safety; it is safety from harm or loss due to spam, viruses, worms, fraud, identity theft, and so on. Many online providers of services, and many online organisations, wish to be perceived as a cyberplaces, where the risk of loss and harm is low. Thus, the 'law' is developing through the interaction of individual service providers and their customers, as well as through the interactions of the members of the various cyber organisations.

Summary Chapter 8 'Legal Issues and Policy of Virtual Identity II'

Regulating by software code is one of the emerging ways of controlling the Internet and also the applications that use various forms of information technology to control the behaviour in the virtual society. The use of virtual identity as a cover for criminal conduct is spotted, same as the status of virtual identity in law of persons and property. We learned how non-discriminating intelligent agents bring down a civilian aircraft, carrying roughly 290 passengers to their last destination. The basis of intellectual property law is not only a-territorial, but is also related to economy. Three variables construct the basis of law and economic analysis: communities, time and individuals. However, keeping the ownership of your own VID is a serious concern, as it will not be understood as property. Finally, knowing the EU 1997 statement that every citizen can choose to remain anonymous online, the legally intangible anonymous identity in the cyberworld seems to be chained. The situation shows the position of the virtual environment in the real world system of rules: free in the cage of the Internet. In the next chapter the legal consequences of the VID will be discussed.

---

[522] TL O'Brien, 'Identity Theft Is Epidemic. Can it be Stopped?' New York Times, nytimes.com, October 24, 2004.

TL O'Brien and S Hansell, 'Barbarians at the Digital Gate', New York Times, nytimes.com, September 19, 2004.

[523] In computer science, a software agent is a piece of software that acts for a user or other programme in a relationship of agency. Such 'action on behalf of' implies the authority to decide when (and if) action is appropriate. The idea is that agents are not strictly invoked for a task, but activate themselves. Related and derived concepts include intelligent agents (in particular exhibiting some aspect of Artificial Intelligence, such as learning and reasoning), autonomous agents (capable of modifying the way in which they achieve their objectives), distributed agents (being executed on physically distinct machines), multi-agent systems (distributed agents that do not have the capabilities to achieve an objective alone and thus must communicate), and mobile agents (agents that can relocate their execution onto different processors).

[524] J Mena, Investigative Data Mining for Security and Criminal Detection. (Butterworth-Heinemann 2003).

[525] Stand alone Computer Game 1997; WWW: <http://cdaccess.com/html/pc/iwar.htm>.

[526] T Busuttil and M Warren, 'Intelligent Agents and Their Information Warfare Implications', 'Proceedings of the 2nd Australian Information Warfare & Security Conference 2001: 'survival in the e-economy', pp 109-118, We-B Centre, School of Management Information Systems (Edith Cowan University, Perth, WA 2001);   and >>

T Busuttil, 'Intelligent Agents and how they can be used within an Information Warfare theatre'. (2006) In; Journal for Information Warfare. JIW Vol.6, Issue 1, 2007-03-31.

9 Legal Consequences of Virtual Identity

'Please make it so complex that I shall be impressed but unable to understand it'. De Bono [527]

Introduction
Manual use of online identities in cyberworld results in a range of virtual identities during the stay. Even when there is no direct user (human) controlled action of your 'second ego' (initiated by yourself) the computers and connected networks will be busy with executing your orders in the virtual world. Virtual identities (VIDs) can also live their own life. The use of autonomous virtual identities (such as the chess (ro)bot as your opponent in a game of chess) is increasing. More and more software contains parts of artificial intelligence that uses the VIDs.
In this chapter the consequences of the virtual identity in the legal system are pointed out. From different views the legal position of the VID is considered. The chapter is completed with an overview of the limits to regulation.

Footboard to Virtual Footprints with Real Foothold
This subtitle discusses the questions raised about the legal consequences of agents.

Artificial Agents
New service oriented applications will deliver and present the user (customer) a package of software, code and content, assembled and personalised by artificial intelligence (AI) 'powered' agents. One can say that somewhere there is always a real person as manager or operator behind the scenes, but the practical situation is that the software is programmed to run automatically and autonomously, and only to make an occasional choice in content management, individual needs (including the presentation requirements for the used devices) and the service offer in order to achieve the quality of service that is agreed between the user and the provider/manufacturer. The AI-generated, modified and removed ad hoc (software-)agents behave themselves during their life as virtual identities.

Viewing the legal aspects of this phenomenon raises questions about the legal consequences. When non-human intelligent agents can live with a virtual identity as a virtual aliens, then who is responsible and liable for their actions in the networks and systems on our earth?
Every investigator will question in advance 'who did what where when and how'? And each court will consider the admissibility, before focusing on the suspects and the evidence. After clearing the competence six participants make their pleas:
- The waver of really autonomic executing virtual identity
- Difference between an identity in the real world and in cyber environment
- Virtual identity in the legal system
- New legal definitions for new virtual identity concepts
- Limits to regulation.

The Waver of Autonomic Executing Agents with Virtual identity
Information and communication technology not only creates new beings in the world, but also affects the conceptual framework we depend on in our understanding of these beings in a fundamental way. As body and mind are closely linked, information technology affects both our mental and our bodily identity. (De Mul) [528]

Identity Generator
All kinds of online and offline virtual identities of software agents are generated by a computer. The

---
[527] E de Bono, 'The horror of the simple', in: 'Six Action Shoes' (1991). WWW: <http://www.edwarddebono.com/newedb1.nsf/>.
[528] J de Mul, 'Networked Identities' In: Michael B. Roetto (ed.), Proceedings Seventh International Symposium on Electronic Art, Rotterdam 1997, p 11-16. WWW: <http://www2.eur.nl/fw/hyper/Artikelen/isea96.htm>.

term 'computer-generated' is not defined further. However in the case of *Express Newspapers v Liverpool Daily Post & Echo* the work under consideration was seen to fall within the scope of being computer-generated. [529] The case seems to suggest that 'a work is computer-generated when the computer is in a sense acting on its own to produce the actual works; i.e. in this case an algorithm was used to select each of the five letter sequences rather than a human making any decision or creative act in each case. Most relevantly the ruling seems to interpret 'arrangements necessary for the creation of the work' as the use of a computer programme, as opposed to the creation of that programme. So, one can argue that the user is part of the 'circuit'.

If we accept the statement that a computer enabled agent with virtual identity can execute autonomously both human and machine related actions it is self-evident that the legal aspects of these actions are related to both human and machine. In the concrete the actions generated by a virtual identity always are related to an origin, such as a trigger of a software programme or a key-press of a keyboard by a human being.

Virtual identities are generated by the software in a machine (computer). So, the fundamental question is about the autonomic executing machine, about the designer of that machine, and about the human who wrote the software code to let process that machine.
In the industrial century, when machines overwhelmed people, there was a broad discussion among scientists about the consequences of the automation. In the spring of 1947, the scientist Norbert Wiener was invited to a congress on harmonic analysis, held in Nancy, France. During this stay in France, Wiener received the offer to write a manuscript on the unifying character of this part of applied mathematics, which was to be found in the study of Brownian motion and in telecommunication engineering. The following summer, back in the United States, Wiener decided to introduce the neologism cybernetics into his scientific theory. In the UK this became the focus for the Ratio Club, a small informal dining club of young psychologists, physiologists, mathematicians and engineers who met to discuss issues in cybernetics.

Wiener pointed out in 1948 the parallel history of the automaton and the human body as 'automata whether in metal or in flesh' [530]. He states in his theory that animals (including humans) and machines are controlled by:
- message,
- noise,
- coding,
- information amount *and*
- feedback.

So, following Wiener's theory – that was broadly accepted by scientists and lead to the cybernetic science – a machine never can be autonomic, because there is always a control, e.g. by message, noise, coding, information amount and feedback. It has not been proved yet that a 'perpetual mobile' [531] really exists. Just as the cyber Walhalla is dependent on the power cord, we can conclude that an autonomic executing machine is somewhere subject to a kind of control. Designing, building and managing machines is done (and controlled) by human beings, even when it is a self-developing machine or self-generating software. Switch off (= controlling) the electricity and the machine stops.

Who Did It ?
In the legal view someone should be responsible for that control, either in design as well as in management or execution. The service of 'my last email' (see page 140) proves that there are at least two designers, one of the content and proposed action, and one of the computer system(s) involved.

---

[529] Express Newspapers plc v Liverpool Daily Post & Echo plc (1985), 1 WLR 1089.

In this case a computer programme was used to generate unique five letter sequences which were printed on 22 million cards as part of a competition called Millionaire of the Month. Council for the defence argues that as there was no human author copyright did not subsist – hence the defendant was free to publish the winning sequence is their newspaper. Whitford J Ruling defined the role of the computer as instrumental, saying 'The computer was no more than a tool' and rejected the defence argument stating 'it would be to suggest that, if you write your work with a pen, it is the pen which is the author of the work rather than the person who drives the pen.' In the ruling the author of the work was adjudged to be the programmer. The ruling is slightly ambiguous as the person adjudged to be the author was both the user of the programme and the programmer.

[530] N Wiener, *Cybernetics: Control and Communication in the Animal and the Machine* (MIT Press, Cambridge 1948).

[531] Perpetual motion refers to a condition in which an object continues to move indefinitely without being driven by an external source of energy. In effect by its very definition, Perpetual Motion is a system wherein the item in question consumes and outputs at least 100% of its energy constantly, sustaining no net loss as a result of the laws of thermodynamics. Using modern terminology, any machine that purports to produce more energy than it uses is a 'perpetual motion machine', although they may not include any moving parts. WWW: <http://en.wikipedia.org/wiki/Perpetual_motion>.

Also, there are at least two managers, one ruling the basic action (the subscriber) and one of the service (the provider). In the execution phase there are several executors, such as the subscriber (who dies), the notary (who confirms the death of the subscriber), the employee of the provider (who executes the message of the notary into a machine action) and all other providers who pass and forward the pre-edited message to its destination. If this specific last email contains a code that results in a new action (such as compiling and executing the distribution of a pro-programmed virus), there is always a link to the origin.

Does someone have a duty to research before executing something unknown? Is Google responsible when their search engine crawls into top secret security information of a government and makes that accessible for terrorists? Are podcasters responsible for the content of outdated podcasts? Looking to the Indymedia case, the answer is "Yes". Nevertheless, it is unclear whether the same result would follow if a podcaster is shown to have 'actual knowledge' that material previously published is false, or a service provider could check-off the personal message. A court might hold a podcaster responsible for failing to remove material known to be false without imposing an affirmative duty to research the continuing validity of previously published statements.[532]

But are a cell phone device manufacturer, a SIM card seller and a mobile telecom provider liable when someone uses (without knowing it) their telecom network and equipment to detonate a bomb? [533] Referring to the Scientology case, the answer is *No*.
It is common that e.g. a lawyer has a professional duty to research and discover the rules and facts, such as a patent agent, journalist and scientist feel the same duty.[534] Or do they face [535] similar professional risk as the gandy dancers of the nineteenth century at the forefront of the westward expansion of the USA-railroads, if they fail to become competent in using the Internet for research?

At least, the ECHR court concluded in such a case, given the lack of procedural fairness and the disproportionate award of damages, that there has been a violation of Article 10 (freedom of speech) of the 'European Convention on Human Rights' [536].

> '*A hacker might hide or 'spoof' his Internet Protocol (IP) address, or might intentionally bounce his communications through many intermediate computers scattered throughout the world before arriving at a target computer. The investigator must then identify all the bounce points to find the location of the hacker, but usually can only trace the hacker back one bounce point at a time. Subpoenas and court orders to each bounce point may be necessary to identify the hacker. Some computer hackers alter the logs upon gaining unauthorised access, thereby hiding the evidence of their crimes.*'   Source: Obstacles to Identifying the Hacker [537]

What is allowed in one state is not accepted in another country. Against the opinion in the US that an anonymous poster should be prosecuted for his/hers/its liable messages, stands the practice in Japan (and other far-east countries) to accept the anonymous posts. (See the '2channel' case, p 152) Opposite, the US is tolerating services such as "mylastemail.com" that let the death act.

Email from the Great Server in the Sky

> '*How would you like to be remembered? Some people can't live without e-mail -- and, apparently, some people can't die without it. Fortunately for those in the latter group, there's mylastemail.com, a new service that promises to deliver your final, heartfelt e-mail messages to your friends and relatives once you have passed on to that big cubicle in the sky. This new, unique web-based service helps you plan for the unexpected and soften the blow of sudden loss. It's actually about life, not death. These emails are prepared by the living, for the living, with a desire to leave messages of hope for those we love and care about.*'   Wired 12-11-2003

---

[532] JP Hermes and SL Gerlovin, 'Podcasting: Keeping It Legal' (2006) WWW: <http://www.ecommercetimes.com/story/53452.html>.

[533] The 2004 Madrid train bombings (also known as 3/11 and -in Spanish- as 11-M consisted of a series of coordinated bombings against the *Cercanías* (commuter train) system of Madrid, Spain on the morning of 11 March 2004 (three days before Spain's general elections). WWW: <http://en.wikipedia.org/wiki/11_March_2004_Madrid_attacks>.

[534] E.g. 'Legal Ethics in the client-lawyer relationship: Lawyers' Manual on Professional Conduct' § 31 (1997) p 201; WWW: <http://www.law.cornell.edu/ethics/al/narr/al_narr_1_01.htm>.

[535] LD MacLachlan, 'Gandy dancers on the Web: How the Internet has raised the bar on lawyers' professional responsibility to research and know the law'(2001) *Georgetown Journal of Legal Ethics*; WWW: <http://findarticles.com/p/articles/mi_qa3975/is_200007/ai_n8903949>.

[536] Steel & Morris v UK government in McLibel case. EHCR 15-02-2005. WWW: <http://www.echr.coe.int/Eng/Press/2005/Feb/ChamberjudgmentSteel&MorrisvUnitedKingdom150205.htm>.

[537] NN,'Tracking a Computer Hacker'; WWW: <http://www.cybercrime.gov/usamay2001_2.htm>.

Looking to New York Times' columnist Art Buchwald's last words 'hi I am Art Buchwald and I just died!' [538], now on video at the Website, one can say that at least the newspaper had its responsibility in checking the message before arranging the broadcast. But until this area of the law is clarified, the cautious approach for publishers and providers who have 'actual' notice that previously published statements are false is to remove the relevant podcasts from circulation, or at least to edit those podcasts in order to remove the false material. The same can be stated for providers who execute the transmission of messages from the deceased.

Who Did What?

Many countries worldwide have enacted criminal laws that prohibit unauthorised access to computers. Each state regulates in its own way. And the nationality of a suspect or convict is also qualifying. Both reservations should keep in mind with the following cases.

> **For example,** the US Indiana State Code 35-43-2-3 states that a person who knowingly or intentionally accesses (1) a computer system, (2) a computer network, or (3) any part of a computer system or computer network without the consent of the owner of the computer system or computer network, or the consent of the owner's licensee, commits computer trespass. A computer system is defined in the code as 'a set of related computer equipment, software, or hardware.'

However in some cases, such as *Intel v Hamidi* the Californian Court of Appeal held that being the owner of the trespassing computer system, it is Intel's request that Hamidi stop sending agitated emails to Intel employees, however, the court declined. [539] After that decision some court cases were passed – with different results – about a hidden action by automatic search-bots of search engines in someone else's computer systems. In a Moscow court case a computer programme for transfer of *Komsomol* members' dues at one domestic enterprise was so developed, that in 1990 deductions had been made not only from the salary of *Komsomol* members, but also from the salary of all the 67 staff members under 28 years old. [540] However, Article 273 of the Russian Criminal Code protects the rights of the owner of the computer system to immunity of the information contained in the system.

In reverse in Japan, with Sony's rootkit-installing CD the accompanying license agreement states that Sony-BMG can install and use backdoors in the copy protection software or media player to 'enforce their rights' against you, at any time, without notice. And Sony-BMG disclaims any liability if this 'self-help' crashes your computer, exposes you to security risks, or any other harm. [541]

There are hardly any court cases where the 'bot' as a mechanically generated virtual identity is judged. Neither are there cases where the owner is judged responsible for a bot-hack into his system, independent of the owner could knowing it or not. Broadly speaking the first step in judgment will be to check any evidence is available that proves that the owner of the used hardware or software really knows if and what the VIDs execute. The second step is the liability to the effects of the self processing virtual identities. In that context the other parties of the 'circuit' are also investigated and 'in charge'.

It can be concluded that even an autonomic looking virtual identity (such as the bot in the software) always has been related to – at least one – human action. But – in legal view – being the perpetrator, supervisor or owner of a computer (-system, -network) is not the same as being accountable, responsible, culpable, and/or liable. Dependent to the law in force and to the specific situation a proof of evil intent or malice aforethought is necessary to be guilty of something.

Legal Difference between an Identity in the Real World and in Cyberworld

Marking if Difference and Exclusion

Identity establishes two relations: similarity and difference (Jenkins)[542]. Hall has summarised this point of view with particular clarity: 'identities' are more the product of the marking of difference and exclusion, than they are the sign of an identical, naturally-constituted unity – an identity in its traditional meaning (that is, an all-inclusive sameness, seamless, without internal differentiation). [543]

---

[538] Ard Buckwald last words. WWW: <http://www.nytimes.com/packages/html/obituaries/BUCHWALD_FEATURE/blocker.html>.

[539] Intel v Hamidi (2002); WWW: <http://cyber.law.harvard.edu/openlaw/intelvhamidi/>.

[540] WWW: <http://www.crime-research.org/analytics/Liability_for_computer_crime_in_Russia/>.

[541] The Legalese Rootkit Sony-BMG's EULA; WWW: <http://www.eff.org/deeplinks/archives/004145.php>.

[542] R Jenkins, *Social Identity* (Routledge, London 1996).

[543] S Hall, *Introduction: Who Needs Identity?* In: S Hall and P du Gay (eds.) 'Questions of Cultural Identity' (Sage 1996).

Above all, and directly contrary to the form in which they are constantly invoked, identities are constructed through, not outside, difference. This entails the radically disturbing recognition that it is only through the relation to the Other, the reaction to what it is not, to precisely what it lacks... that the positive meaning of any term - and thus its 'identity' – can be constructed...identities can function as points of identification and attachment only because of their capacity to exclude...'

Clarke & Saraga argue 'How we view the conditions and consequences of social policies depends very heavily on how we understand the society in which such policies are formulated and implemented. At the centre of this problem is the question of how we make sense of patterns of difference between individuals and groups in society.' [544]

We might ask a number of questions about the issue of difference:
- What sorts of differences are visible in our society?
- What sorts of differences have consequences in our society?
- What do we do about such differences?
- Where do these differences come from? '.

As long as the identity of an individual is influenced by her/his society the equalities and differences are essential in the exposure of someone's identity.

Woodward argued in an Open University text that: 'old certainties no longer obtain and social, political and economic changes both globally and locally have led to the breakdown of previously stable group membership. [545] Identities in the contemporary world derive from a multitude of sources – from nationality, ethnicity, social class, community, gender, sexuality – , sources which may conflict in the construction of identity positions and lead to contradictory fragmentary identities.'

Wenjing states that with the advent of globalisation, the movement of immigrants has crossed the borders and the question of identity has been more and more salient in the modern society. [546] Struggling in different cultural frames, Diasporas' cultural identities are constructed through the interaction between what Hall called similarity and difference in the process of displacement and relocation. Scattered across the western hemisphere, the new Chinese Diaspora are gathering in the virtual communities to grab news, to comment on issues, and to negotiate their identities in an alien culture. The negotiation of different cultural values and practices results in the fragmented and hybridised cultural identity among them in virtual communities. This exploration of interpersonal experiences also shows that the symbiotic relationship between the physical community and the virtual community forges and strengthens such discursive cultural identity.

Rheingold states that 'Virtual communities are social aggregations that emerge from the Net when enough people carry on (...) public discussions long enough, with sufficient human feeling, to form webs of personal relationships in cyberspace.' [547] He also argues that the use of nicknames is based on an artificial but stable identity. 'You never know who the person is behind a nickname but you can be quite well sure that the one who was using a nickname yesterday is mostly the same as the one who uses that nickname today.' Rheingold is sceptical about the truth on Internet: 'You can be fooled about people in cyberspace, behind the cloak of words. (...) computer mediated communications provide new ways to fool people, and the most obvious identity swindles will die out when enough people learn to use the medium critically'.

Clarke concluded that we have developed a digital identity besides our real world identity. [548]This digital identity is a model, based on information that exists about us in the virtual world and that is used as replacement of our real world identity.

Bechar-Israeli researched the behaviour behind nicknames and discovered that nicknames say something about the person behind the name. [549] Near 45% of the researched nicknames could be somehow or other, by hook or by crook, related to the person (al virtue).

---

[544] R Clarke and E Saraga, 'Introduction', in: E Saraga (ed) 'Embodying the Social: Constructions of Difference' (Routledge, London 1998)

[545] K Woodward, *Concepts of Identity and Difference*, in K Woodward (ed) 'Identity and Difference' (Sage/OU, 1997).

[546] X Wenjing, 'Virtual space, real identity: Exploring cultural identity of Chinese Diaspora in virtual community'. *Telematics and Informatics* Volume 22, Issue 4 (November 2005).

[547] H Rheingold, *The virtual community: Homesteading on the electronic frontier.* (Addison-Wesley, Reading 1993).

[548] R Clarke, 'The digital persona and its application to data surveillance', *The Information Society*, v10, iss. 2, June 4 (1994).

A nickname can express someone's looks (belladonna), someone's character (shy dude), or someone's job (director). Bechar-Israeli signed that in only 7,8% of the cases people used their own name. He also measured that during the time of the research persons seldom changed their nickname.
Turkle concluded in her book 'Life on the Screen: Identity in the Age of the Internet' that anonymous people online always show something of their online characteristics in their real life, and reversed. [550]
Virtuality is always related to - and interacting with - real actual appearance, argues Van den Boomen after her hands-on experience in virtual communities. [551]

These previous statements position "identity" (in both the physical community and the virtual community) not anymore as something individual and outlaw or 'above the law', but brings the changes in and expressions of identities within the society, and in the (system of) rules of that society.

In such a context there will be no theoretical difference between an identity in the real world and that in the cyberworld but only a practical one: the medium.
Just as the numerous ways laws come into everyone's lives, the media on earth also are ruled by laws. [552] Law has been defined as a 'system of rules' [553], as an 'interpretive concept'[554] to achieve justice, as an 'authority' [555] to mediate people's interests, and even as 'the command of a sovereign, backed by the threat of a sanction' [556]

Virtual Identity in the Legal System

In this subchapter the opinion is discussed that the activities of VID's are exempted from the basic legal principles in the real world. It is considered also if the use of technology code is a way to achieve desired behaviour to accord to the norms.

According to the Conditions of Our World

Code, Code or Code

> '... You do not know our culture, our ethics, or the unwritten codes that already provide our society more order than could be obtained by any of your impositions. You claim there are problems among us that you need to solve. You use this claim as an excuse to invade our precincts. Many of these problems don't exist. Where there are real conflicts, where there are wrongs, we will identify them and address them by our means. We are forming our own Social Contract. This governance will arise according to the conditions of our world, not yours. Our world is different.'
>
> Quotes from 'A Declaration of the Independence of Cyberspace' by John Perry Barlow [431]

In the beginning the Internet nerds said 'we make law ourselves' as did the colonists. Indeed, in the early years the behavioural code seemed to be sufficient, and a ban for some days from the community was generally enough to correct the misbehaviour.
Lists of IP addresses that were suspended upon expulsion were distributed in order to lock out their excess to several parts of cyberspace. The banishment seemed to be an effective way to let the cyberians 'accord to the conditions of our world'.

---

[549] H Bechar-Israeli, 'From WWW: <Bonehead> to WWW: <cLoNehEAd>: Nicknames, Play, and Identity on Internet Relay Chat', *Journal of Computer-Mediated Communication*, vol. 1, issue 2, 1997.

[550] S Turkle, *Life on the Screen: Identity in the Age of the Internet*, (Simon and Schuster, 1997).

[551] M van den Boomen, *Leven op het Net* (Instituut voor Publiek en Politiek, Amsterdam 2000).

[552] The numerous ways law might be thought of reflects the numerous ways law comes into everyone's lives. Contract law governs everything from buying a bus ticket, to obligations in the workplace. When buying or renting a house, property law defines people's rights and duties towards the bank, or landlord. When earning pensions, trust law protects savings. Tort law allows claims for compensation when someone or their property is harmed. But if the harm is criminalised, and the act is intentional, then criminal law ensures that the perpetrator is removed from society. WWW: <http://en.wikipedia.org/wiki/Law>.

[553] HLA Hart, *The Concept of Law*. (Oxford University Press, 1961) p 12 ff.

[554] R Dworkin, *Law's Empire*. (Harvard University Press, 1986) pp 92-96.

[555] J Raz, *The Authority of Law*. (Oxford University Press, 1979) pp 26-33.

[556] J Austin, *The Providence of Jurisprudence Determined*. [1831] (Murray, London 1832. Reprint. The Legal Classics Library, Birmingham 1984) (Cambridge University Press; Nw edn 1995) p XXI.

In the early Greek history corporal punishment was common, in all classes of society. The most extreme corporal punishment was the death penalty. Plato objected to this excess of the penal code, because he argued that punishment should be considered after learning the social code and that was no longer possible for the decedent. [557] The same line of thought is applicable for virtual misbehaviour. When 'banned' from a Website or community (e.g. by blocking the IP address of the abuser) the excommunicated person will find a new place to impose upon. Depending on whether they learned the lesson or not and their attitude, the behaviour in the new environment will improve or remain the same as in past.

But the ideal of 'our own Social Contract' was not sufficient to stop the vandals and criminals in their attempts of identity theft and fraud. After some time the traditional laws, mainly focussed on the territories of a country, "rescued" and ruled several illegal actions through the telecom connection that started or passed the territories of that country. Even when more countries were involved, the courts are creative in their condemnation. Comparing with cases in the real world, deciding in analogy with similar best practices in the physical environment and some 'a contrary' reasoning are helpful to correct the human behaviour on the Internet by jurisprudence.

The EU states that 'there is a clear consensus that activity on the Internet cannot be exempted from the basic legal principles that are applied elsewhere. [558] The Internet is not an anarchic ghetto where society's rules do not apply. Equally though, the ability of governments and public authorities to restrict the rights of individuals and monitor potential unlawful behaviour should be no greater on the Internet than it is in the outside, offline world. The requirement that restrictions to fundamental rights and freedoms be properly justified, necessary and proportional in view of other public policy objectives, must also apply in cyberspace. This principle of treating the Internet no more or less favourably than older technologies is reflected both in the introduction to the Commission's Communication on Illegal and Harmful Content on the Internet which states 'what is illegal offline remains illegal online'.' With analogy to this statement it can be expected that codes offline will become online.

The Range of the Law

Laws addressing computer source code and hacking `tools' are often not applicable. As evidenced by the past cases involving hackers, jurisdiction can be a problem. In many cases, the alleged perpetrator physically resided in the US, the system he/she reportedly attacked was located in Europe, and the used 'hacking' network was located all over the world. Next moment the locations have changed to Russia and Australia. The question of jurisdiction has made prosecution impossible, states Cook. [559]

Laws concerning virtual identities have problems due to their lack of enforceability, jurisdiction and the matter of recovery, argues Gordon. [560] The nature of the methods of creating virtual identities tends to hamper any real assessment of exactly how much information is being exchanged and by which identity. Of course there are ample mechanisms for monitoring information exchanges, but we need to be concerned with various policies (legal, ethical and technological) when we consider monitoring communications to ensure their 'acceptability'.

The jurisprudence in countries worldwide indicates that courts found that information distributors are not strictly liable for damage caused by distribution of misinformation. However, decisions also have held that distributors of products can be held strictly liable for the results of reliance on misinformation contained in the product. (Cook)
While laws are still evolving and no one knows for sure what the end result will be, it shows realism that computer (network) administrators and commercial system owners will face possible liabilities for actions of their users. They may have a responsibility ethically and may have eventually legally to know what is going on within their systems, but one cannot ignore the obvious gap between what

---

[557] TJ Saunders, Plato's Penal Code: Tradition, Controversy, and Reform in Greek Penology (1991). A detailed exposition of the emergence of the concept of publicly controlled, rationally calculated, and socially directed punishment in the period between Homer and Plato. Plato advanced the most radical of the philosophical formulations of the concept of punishment in his Laws, arguing that punishment is or should be utilitarian and strictly reformative. A serious debate ensued in the 5th century over the opposition by philosophers to popular judicial assumptions and their arguments gradually gained ground.

[558] Ministerial Conference, Bonn, 6-8 July 1997. *Anonymity on the Internet.* Recommendation 3/97. WWW: <http//ec.europa.eu/justice_home/fsj/privacy/docs/wpdocs/1997/wp6_en.pdf>.

[559] WJ Cook, Network Traffic Liability: op-ed for AAAS Invitational Conference on Technical, Ethical and Legal Aspects of Computer and Network Use and Abuse (1993).

[560] S Gordon, 'Technologically Enabled Crime'(1999) WWW: <http//www.research.ibm.com/antivirus/SciPapers/Gordon/Crime.html>.

a system should enforce and what it is actually expected to enforce. The same unpredictably arises with the enforcement of the social policies and mores that exist in any given environment. (Neumann) [561]

More and more the use of technology code is a way to achieve a desired behaviour. Social networking Website "MySpace.com" announced last year that it was developing technologies to help block convicted sex offenders from using its Website. [562] Competitors such as Google (YouTube) will follow. Both ethical and legal solutions to some of the problems discussed in this chapter are worth considering. However, both have limitations, and need to be used in a cooperative or multidisciplinary approach, e.g. methods to address the problems.

The Virtual Party in the Circuit

Who is a 'Party to the Communication' in a network (ab)use? In the earlier paragraphs of this chapter the 'circuit' reasoning was discussed and also shown were some cases in which was cleared that a person hanging around a suspected scene would also be suspect.

US law [563] permits any 'person' who is a 'party to the communication' to consent to monitoring of that communication. In the case of wire communications, a 'party to the communication' is usually easy to identify. For example, either conversant in a two-way telephone conversation is a party to the communication, but in a computer network environment, in contrast, the simple framework of a two-way communication between two parties breaks down.[564]

Although people of different nationalities can be treated several ways in dissimilar countries, more and more of the courts worldwide are deciding that a suspect, operating in a computer network, can be judged as 'party in that circuit', similar to cases in the 'real' world.

My Thinking VID

People often express their thoughts in a virtual environment as text or images, more than they would do in real life. Cyberspace is experienced as a part of a private home, the same as Usenet in the earlier days. The Internet medium makes it possible (and by the experienced anonymity often attractive) to reflect and test one's thoughts by trying out various roles, identities and environments. Using thought-making Websites such as voting advisors people can make up their mind for the next election.[565] Following the discussions on the Internet about the interest of domestic intelligence services this kind of free thought reflection and expression could be seen as the 'speech' of the person 'behind' the VID. The evolving Web 2.0 with far-reaching individual identity management enables us to link the virtual identities not only to IP addresses but also to the identity of individuals.

My Stolen VID

Worldwide the theft of virtual identities is reported. In most cases it concerns identity theft by hacking or threat. In many situations it is a kind of pilferage but is some remarkable cases the online identity of the virtual identity was stolen, sold, killed or despoiled of the valuable avatar, game level or score points. Even at the highest rating auction sites such as eBay and Amazon scamps and scoundrels perpetrate a crime. The crime is simple: having taken over an identity (like having the door key of a building) enables them to take out /over the valuable digital goods, and to subsequently disappear with the loot.
Having no virtual hands-on experience, the police do not suspect foul play when the victim reports the crime.

Jurisprudence in a lot of countries shows that for this kind of crime traditional laws can be used, as it is a virtual variable of an economic delict, a criminal offence or a violation of the private law. As long as all tracks and traces are kept and the forensic police (can and will) investigate the log files with the service operator and the IP routing used, most of the reports can be brought to court.

---

[561] PG Neumann, *Limitations of Computer-Communications Technology*, AAAS Invitational Conference on Legal, Ethical and Technological Aspects of Computer and Network Use and Abuse (1993).

[562] 'MySpace to Block Sex Offenders' BBC News, 07-12-2006. WWW: <http://news.bbc.co.uk/2/hi/technology/6216736.stm>.

[563] 'Searching and Seizing Computers and Obtaining Electronic Evidence in Criminal Investigations' (2001) Computer Crime and Intellectual Property Section; Criminal Division; US Department Justice. Sections 2511(2)(c) and (d). WWW: <http://www.dojgov.net/techno_confiscation.htm>.

[564] See, e.g., United States v. Davis, 1 F.3d 1014, 1015 (10th Cir. 1993).

[565] The Dutch Data Protection Authority WWW: <http://www.dutchdpa.nl/> started research to the use of the interactive election vote advisor "Stemwijzer". WWW: <http://www.stemwijzer.nl/> related to the IP addresses of the site visitors. More examples: WWW: <http://www.politix.nl/projects/nieuwekieswijzer/> and: WWW: <http//www.fnvkiestpartij.nl/fnv-kieswijzer-2006.60.html>.

Before going to court, the service provider involved can also be helpful. For instance eBay and PayPal have a Fraud Investigations Team (FIT) in order to promote the safe use of its platforms and to collaborate with local, state, federal and international law to enforce policies, prosecute fraudsters, and help keep the community safe. Disputes about the bid, goods or payment are handled by professional mediators in a fair way.

Given the booming flood of online transactions (and the derivative of conflicts) it is to be expected that more and more commercial and peer group parties will mediate in virtual transactions. As opposed to the real punishments in the real world, this virtual law without order has a lead: it can virtually victimise in an effective and broadly-based way.

My Naked VID

Not every phone call is considered as pleasant and welcome, same as email. In the past, all over the world, there have been disputes and court cases concerning identity control and prevention against anonymous telecommunication acts. In the meantime, most countries have a statute barring telephone harassment in the form of anonymous, repeated, stalking, obscene, misleading, or non-consensual calls, even when they are automatically dialled or pre-recorded.[566] Also when free and anonymous receiving of broadcasts changes identity controlled reception (e.g. radio and television over IP) [567] the idea of free communication diffuses. Significant is that where countries are limiting the anonymous identities by law, technology whiz kids create new kinds of anonymous virtual effects that result in even more of the same. Perhaps that is the relation to the observation that new media are regulated more than the older ones.

Nowadays the information technology presents us multimedia varieties of automatic and pre-recorded interactive communication contacts, using at least two human senses, increasing in the next decennium to almost all senses. Weinstein stated: [568]

Taking Liberties With Our Freedom

'Since the events of 9/11/2001 in the USA, a range of legislation detrimental to fundamental freedoms and privacy rights has been rammed into US law, without any assurance that safety will improve as a result. While it has some positive aspects, the US Homeland Security Act [569] is also full of worrisome surprises for U.S. citizens concerned about their freedoms, particularly when combined with the USA Patriot Act. [570] Among various alarming provisions, the law opens up enormous avenues for monitoring Internet communications, without even after-the-fact notifications. Virtually any government agency at any level can initiate surveillance in flimsy grounds. No subpoenas or court oversight are required. Homeland Security Act creates new and very broad exemptions from the Freedom of Information Act.' [571]

The main difference between (online) anonymity and pseudonymity is that – while in anonymity the identity is not known in pseudonymity – a separate persistent virtual identity exists but that it cannot be linked to a physical person or organisation. Law enforcement officials keep expressing their opposition to online anonymity and pseudonymity, which they view as an open invitation for criminals and outlaws who wish to disguise their identities. Therefore, the officials often call for an identity management infrastructure in tandem with a secure digital national identity document.

---

[566] E.g. the US Communications Act of 1934, 47 U.S.C. s 223(a) (1988), bars such calls in interstate or foreign communications. In addition, the Telephone Consumer Protection Act of 1991, 47 U.S.C. s 227(b), bars automatically dialed or prerecorded telephone calls. (Harvard Law Review March, 1994). WWW: <http//www-swiss.ai.mit.edu/6095/articles/message-in-the-medium.txt>.

[567] E.g. US Court case FCC vs. Midwest Video Corp., 440 U.S. 689, 700-02 (1979) holding that public access requirements for cable television systems imposed common carrier obligations because they deprived the operator of any control over the content transmitted and the identity of users. (*Harvard Law Review*, March, 1994). WWW: <http//www-swiss.ai.mit.edu/6095/articles/message-in-the-medium.txt>.

[568] L Weinstein, 'Taking Liberties With Our Freedom' *Wired News* (2002); WWW: <http://www.wired.com/news/politics/0,1283,56600,00.html>.

[569] Homeland Security Act of 2001, H.R. 5005, Sec.1005: Inventory of Information Systems: The identification of information systems in an inventory under this subsection shall include an identification of the interfaces between each such system and all other systems or networks, including those not operated by or under the control of the agency. WWW: <http://www.pfir.org/2002-hr5005>.

[570] To deter and punish terrorist acts in the United States and around the world, to enhance law enforcement investigatory tools, and for other purposes, the H.R.3162 Act of 2001(USA Patriot Act) passed the House and Senate of the United States of America to 'Uniting and Strengthening America by Providing Appropriate Tools Required to Intercept and Obstruct Terrorism' by e.g. verification of identification (sec.326), identification of originators of wire transfers (sec.328), fingerprint identification system (sec.405), DNA-identification (sec.528), identification of foreign intelligence (sec.908), et cetera; WWW: <http://thomas.loc.gov/cgi-bin/query/D?c107:4:./temp/~c107l8tcjC>.

[571] The Freedom of Information Act of 1996, 5 USC § 552, As Amended By Public Law No. 104-231, 110 Stat 3048. WWW: <http://www.usdoj.gov/oip/foia_updates/Vol_XVII_4/page2.htm>.

This digital hobble would irrevocably tie an online identity to a person's legal identity. Online civil rights advocates, in contrast, argue that there is no need for a privacy-invasive system because technological solutions, such as reputation management systems, are already sufficient and are expected to grow in their sophistication and utility. Can laws be the solution to controlling the virtual identities?

> **Europe seeks to tighten some online laws** [572]
>
> FRANKFURT, Germany --Some European countries are proposing outlawing the use of fake information to open email accounts or set up Web sites, a move intended to help terror investigations but which could face resistance on a privacy-conscious continent.
>
> The German and Dutch governments have taken the lead on the proposals, crafting legislation that would make it illegal to provide false information to Internet service providers and require phone companies to save detailed records on customer usage.
>
> The aim, analysts say, is to make it easier for law enforcement to access information when they investigate crimes or terrorist attacks. But Europeans have long cherished their privacy, railing against measures that would see personal information stored for commercial use or government examination.
>
> 'The people of Europe have a long record of fighting for their personal freedom, and are unlikely to accept such regulations being imposed upon them,' said Graham Cluley, a senior technology consultant with the London-based consulting group Sophos. 'No one disagrees with the need to take decisive action against terrorism and organised crime, but to introduce such restrictive surveillance on the general public and Internet companies - without proper safeguards in place - seems positively Orwellian,' he said.
>
> Look Christian, 42, who works at an Internet cafe in Berlin, said it's his business - not the government's - if he wants to set up an anonymous email account. 'I understand that the police might want to hunt people down on the Internet, and I wish them luck, but it's not going to happen through anonymous Internet accounts,' he said.
>
> The Germans and Dutch are moving well ahead of a 2009 EU deadline to implement its Data Retention directive, which calls for storing names and addresses of Internet subscribers, including those who use Web-based email accounts. Countries will be able to decide individually how long to keep the information on file, within a range of six to 24 months.
>
> But some details of the new proposals have yet to be worked out. For instance, most of the major Web-based email providers such as Google's Gmail or Microsoft's Hotmail require nothing more than a user name and a password to set up an account. Real names and addresses are not requested.
>
> Simon Hania, technical director of the Dutch Internet service provider XS4ALL, also points out that knowing who pays the bill for an ISP will not necessarily allow police to know who uses a personal email account.' For each and every email address we at least know who is paying for it, which is not necessarily to say who is actually using it,' he said.

My Suspected VID

A (natural or legal) person is only suspected at the moment that there are reasons to call suspicion. (Code of Criminal Procedure; ID law) [573] To be suspected, there should be relevant tracks. Based on facts related to the person himself, should there be at least an idea that something should have been done that is not allowed in our society. An investigator will always question 'who did what where at what time and how'?

So, there should be a notion of presumptive suspicion, like starting facts that lead to circumstantial evidence. The leading question is 'how much technical based indications do I have to gather to point someone out as suspect?' And, not unimportant, the accumulation of the collected facts shall comply with technical standards.

In situations of preventive or subversive investigation both investigator and judge will consider 'which means will be my choice?' To execute this, four considerations rule: 1) legibility; 2) finality

---

[572] M Moore, 24-02-2007. WWW: <http://www.examiner.com/a-583469~Europe_Seeks_to_Tighten_Some_ Online_ Laws.html>.

[573] Identification law is changing in this reserve, as today in countries police can stop someone and ask for his/hers identity. See: 'Convention for the Protection of Human Rights and Fundamental Freedoms' 04-11-1950, Trb. 1951, 154 (Article 5[1] – Right to liberty and security); WWW: <http//conventions.coe.int/Treaty/en/Treaties/Word/005.doc>.

(justified purpose); 3) subsidiality (only allowed if it supports others); and 4) proportionality. For example: an investigation or punishment could be disproportional and on balance may need to be adapted. But what is 'disproportional' in a virtual environment? Who can decide what is justified in a worldwide MMORPG?

Online Penalties

In his 'Laws' and also elsewhere, Plato advanced a radical new penology, sharply different from any penology practised or advocated down to his day. Plato catalogued (in his penal code at the beginning of Laws, Book 5) the ways in which men 'dishonour' their souls. The worst penalty that can be incurred by the wrong-doer is that he is cut off from the society of the good and incorporated into that of the bad and completely assimilated into them. In general, the orators who pleaded before the courts which enforced the penal code of the law of classical Athens betrayed scant interest in crime-specific punishments of any kind at all. Punishment seemed to be a part of Athenian life, at any rate of life as affected by the courts and the legal code. Saunders argues that the 'heresy' law in Plato's model penal code was remarkably sophisticated. [574] Unlike Athenian legislation, Plato's law centres on the psychological state of the offender and carefully distinguishes between heretical belief and impious actions and the damage caused by expert and non-expert magicians.
The similarity with the virtual world is that online and virtual identities also can be excommunicated by means of ignoring their VID or blocking the IP address of the related human 'puppeteer'. This kind of exclusion is felt as a heavy punishment in the virtual community. Sometimes, is seems not to be effective – as the punished identity masks him/herself and returns – but as soon as the outlaw behaviour appears again, the convict is unmasked and the punishment is repeated until the behaviour ends. In extreme situations the referenced Internet service provider is informed, directly via the Abuse Prevention Systems and the Realtime Blackhole (DNSBL)[575].

Cybercrime, What's Next?

The comparison between the real and the virtual shows that new media result in new types of crime. See examples in figure 33, page 96.
Whiz kids, programmers, criminals and investigators are using the same tools to assist them. Such tools can be used also to cause an anonymous 'Denial of Service' by a worldwide botnet. (see: p 84) With the new media technologies, the choice of law issues (such as domicile) may become crucial as the exploitation of identity can easily and quickly occur at national levels in ambiguous jurisdictions. [576] A well communicated consensus on the 'virtual' law would promote certainty and reduce forum shopping.

Limits to Regulation

This subtitle discussed that we are not able to prevent and solve every criminal attempt in a theoretical or technical way, thus we have to search for a balance between risk and regulation. The need is pointed out that open source of all kinds of legal code.

The Code and the Coders

Virtuality is in a human's fantasy. Outlaw behaviour is part of life. One cannot prevent abuse or thieving. One cannot fight windmills, nor NPC's. Abuse, cheating, misconduct and criminality are

---

[574] TJ Saunders, *Plato on the Treatment of Heretics*; in: L Foxhall and ADE Lewis (eds.), 'Greek Law in its Political Setting: Justifications not Justice'. (Clarendon Press Oxford 1996) p 172 ff. See also:
TJ Saunders, *Plato's Penal Code* (Clarendon Press Oxford 1994/2001).

[575] A DNSBL is a means by which an Internet site may publish a list of IP addresses that some people may want to avoid and in a format which can be easily queried by computer programmes on the Internet.

[576] The US Telecommunications Act of 1996 (Pub. LA. No. 104-104, 110 Stat. 56) {WWW: <http://www.fcc.gov/telecom.html>} has prohibited the making of telephone calls or the utilisation of telecommunications devices 'without disclosing (one's) identity to annoy, abuse, threaten or harass any person at the called number or who receives the communications.' The same law also has been clear that the term 'telecommunications device...does not include an interactive computer service.' This means this law has not been aimed at Internet communications. But... to prevent cyber stalking the constitutionally protected rights eroded by the Violence Against Women and Department of Justice Reauthorisation Act 2005 (HR 3402), {WWW: <http://www.govtrack.us/congress/bill.xpd?bill=h109-3402>} which brings the reach of the above-quoted text home to the Internet. The provision in question applies to 'any device or software that can be used to originate telecommunications or other types of communications that are transmitted, in whole or in part, by Internet.'

human and part of our society, and we have to tolerate and live with it. We are not able to prevent and solve every criminal attempt in a theoretical or technical way. Every person runs the risk in our society. Not that we don't have to do something against criminality but the policy should be realistic, not a frenetic searching for inarticulated conditions and impossible solutions.

One side that we have to consider is that – even though we should have cybercops, virtual investigators and artificial judges – we could not regulate all, as neither could their equivalent in the real world.

The opposite is that current code (by the individual conditions of use for using the Internet connection and/or the entering some virtual communities) do not guarantee decent usage in a satisfactory manner. Second, it is commerce driven abuse since you can do more once you pay more. The virtual world is fake but some code can help to add value.

The current architecture of cyberspace is build around a network of distributed systems consisting of billions of computers in which each individual machine has its own code and sovereignty. Even when there should be a central government to control all these computer slaves, the performance of the systems would not be able to process all these code exchanges. Of course, the architecture of the Internet theoretically can change, but not indefinitely as some of the regulations by architecture have limits due to structure of code and the tradition of sharing among coders.

The Coders

An essential component of cyberspace, the people, strengthens the limits computer systems have to process the code.

First, code writing itself cannot be strictly controlled because there are many individuals writing code. Even kids know how to read and write HTML code. Copying code is a basic operation that the owner of any computer can execute. There is no way to prevent copying data if it is in memory. This code can be changed of course to achieve a trusted worldwide networked system, but such a trusted system cannot be implemented without replacing every bus and processor in the world and then making each copy operation a database transaction over the network. Both are not realistic because this will raise the costs and slowdown the processing extremely.

Second, should there be any governmental or commercial obstruction at any moment, the coders can find a way to construct a new network with the desired freedoms and share that knowledge. Coders have a tradition of sharing their code with the non-coders. Ordinary users usually benefit from such circumvention. In almost every country free software circulates to decode encrypted code. It is not because the world is full of hackers; it is because there is great a demand for such code. History learns that there is a human driven continuation of the aversion and resistance among coders against codes that impose undesirable regulation such as privacy theft by companies and overpriced copy protected software. [577]

Summary Chapter 9: 'Legal Consequences of Virtual Identity'

This chapter views the legal aspects of the virtual identity and raises questions concerning the legal consequences of autonomous operating VIDs. It surveys who could be responsible and/or liable for their actions in systems and networks on our earth.

Non-human virtual identities that live their lives as virtual aliens will make every investigator panic. From the conceptual level to some significant details the question of jurisdiction is explored. Main issues as the position of the virtual environment in the real world system of rules, and the difference between an identity in the real world and in the cyberworld are pointed out in depth. The faltering of real autonomy executing virtual identity is worked out and examples of unwritten law and common code are presented. Topics as virtual identity in the legal system and the new legal definitions for the new virtual identity concepts help in understanding the phenomenon.

Regarding sanctions the proposition is put forward that – even though a real judge will punish – the penalty can be more effectively executed in the virtual world.

Next, the final chapter will guide us to the future and it discusses the way forward in a world that is ruled by code, and in which the focus seems to be very much pointed to national law making and very less so to arbitration and other ways of solving disputes.

---

[577] M Fetscherin, 'Evaluating Consumer Acceptance for Protected Digital Content', in: *Digital Rights Management* (2003) ISSN 0302-9743.

10 The Way Forward

The code you need to copy is CaSe SeNsItIvE and is required to prevent spam. [578]

Introduction
Millions of people around the world duplicate their identity in order to inhabit virtual words: cyber societies and multiplayer online games where characters can live, love, discuss, trade, mask, cheat, steal, and have every possible kind of adventure and transaction. Far more complicated and sophisticated than early discussion groups and video games, people now spend countless hours in virtual universes like "SecondLife" and "World of Warcraft" creating new identities, falling in love, building cities, making rules, and breaking them.
This chapter discusses the way forward in a world that is ruled by code, and in which the focus seems to be very much pointed toward national law making and very less so to arbitration and other ways of solving disputes. Some conclusions and recommendations close this dissertation.

The Code that Rocks the Cradle Rules the World
As digital simulated worlds become increasingly powerful and lifelike, people will employ them for countless real-world purposes, including commerce, education, medicine, sexual pleasure, law enforcement, military training and terrorism. In the meantime, by communications and transactions, virtual worlds are being fully integrated into our real-world society and systems. Disputes utilising the Internet were only expected to be a sense of deliberation and understanding of our democracy. By means of virtual communities people can collectively argue and think about the political choices they have to make, such as juries in US courts. Then people may give the very decisions that we need. Inevitably, sooner or later, real-world law will regulate them. But what rules should govern virtual communities? Will it be by our regular legal system or should they be treated as separate jurisdictions with their own forms of dispute resolution? Referring to the history and observing the power of computing code a synergy of all kinds of regulating codes seem to have the best chance.

New Legal Definitions for New Virtual Identity Concepts

Virtual Law
As the Virtual Identity is the representation of a real identity in a virtual environment – where it can exist independently from human control and can (inter)act autonomously in a real electronic system as well as in a virtual global network – the most natural solution should be to embed the virtuality in the common legal system. Thus, when will the time come that the Internet is embedded in traditional codes? Or is the territorial dissimilarity in sovereignty reason enough to find another suitable solution, e.g. regulation by software code. Or is it a combination of all codes?
As it did in domestic and international law, the emerging custom in cyberspace will influence as 'social code' the evolving legal and regulatory framework, but should not be relied upon to become "the law". There can be no guarantee that the practices of the early adopters or the Internet community in general will meet the genuine needs of society at large – especially when that society is global and the use of the Internet as commodity is diverted to practical applications.
Just as the airspace and the extraterritorial waters – where people are passing though in planes and ships – the Internet is a (i.c. electronic) medium that disregards geographical boundaries and needs international cooperation.
Lessig argues that these circumstances throw the law into disarray by creating an entirely new phenomenon that needs to become the subject of clear legal rules but that cannot be governed, satisfactorily, by any current territorially-based sovereign. [579]

---
[578] Instruction: 'Submit the displayed code to post your comment at a blog site'; WWW: <http://blogger.XS4ALL.nl>.

As with the open sea and free air as stated before, this points to a global solution that provides globally enforced governance rules managed by a global institution. WSIS (see p 102) has drawn attention to the need for a solution that reflects certain key principles: transparency, inclusiveness and democracy. It could be that – similar to the WTO and its role in international trade – such an institute will rule the code in cyberspace.
In the meantime, cooperation and treaties between countries, grouped by regions, will make progress in spreading the best practices and discourage the bad habits. The international environmental laws, e.g. on toxic and hazardous chemicals, are a good example of a united 'brains and balls'.
There have been more governance mechanisms and arrangements proposed or created for these kind of 'superterrestrial' issues. For spam, for example, the "OECD toolkit" is a combination of proposals for legal, technical and operational action against spam, which is slow to uptake and implement in each country, and also dependent on enacting legislation of appropriate scope and quality. [580]
Furthermore, regarding its enforcement, under authorities that will be first and foremost national, not global. The toolkit is updated daily. As yet, no global governance arrangement has been proposed that can be of a higher level of effectiveness for dealing with this problem.

The WGIG points out in its Internet governance book (see p 125) that knowledgeable experts have made findings in issues such as cybercrime, where the global governance proposals that could have some effect in fighting these scourges have to begin with the recognition that very little crime on the Internet is strictly cybercrime (more of it is common crime committed through different communication means including the Internet). Pisanty [581] argues in this book that fighting crime that uses the Internet is more "raining" lawyers, judges, and law enforcement officials in tasks that can be as basic as properly seizing physical evidence of crimes, establishing specialised prosecutors endowed with appropriate tools and staff, and so on, again mostly under national jurisdictions. He wonders if this is why institutions that perform effective global governance against spam or cybercrime do not exist at present. The purpose of studying why such institutions do not exist, though, should be clear: the global governance arrangements for the Internet that have succeeded have first and foremost been directed to solving specific problems, and they have succeeded when they have advanced the solution of the problem each has been charged with, involving all relevant stakeholders in a properly organised fashion.

The EU (RAPID) [363] stimulates the legal and regulatory requirements as well as privacy enhancing technology (PET) embedded in information systems – now frequently seen as 'business inhibitors' – as 'business enablers' in this networked world. Also, in US and Asia-Pacific traces of an emerging online identity regulation can be found. Perhaps other countries will follow soon, albeit that the privacy-embedded 'Seven Laws of Identity' will not satisfy every country. The Web will not be what it was.
Another variant is to create international virtual law, such as the current international private law. Or are we already in the stage that (established) norms in the virtual world overrule the law in force? The case-study – how ranchers resolve disputes over cattle trespass, fencing , and negligence – in the partly empiric research of Ellickson referring a dispute-solution for a dispute about strayed cattle in Shasta County, California, shows that social norms can overrule (ignore) the law in force. [582] Why shouldn't we compile one code from all usable codes, including the mechanism of arbitration and dispute-solution?

Customary Law and Arbitrage
Interjurisdictional competition (and court shopping) can occur between legal systems attempting to monopolise law making and enforcement. As Benson argues, 'this is another obvious benefit of inter-jurisdictional competition, and importantly, there is another source of competition as well: customary law can be produced and supported by institutions which are not attempting to monopolise law, and these legal systems offer an alternative to escape the jurisdictions of those who seek such monopolies. Customary law provides a mechanism for avoiding the politicised law of nation states, and the choice of customary law's jurisdictions in cyberspace is becoming increasingly

---

[579] L Lessig, presentation @ Internet Law Program 2003, Rio de Janeiro, Brazil (2003); WWW: <http://cyber.law.harvard.edu/ilaw/brazil03/participants/lessig1.pdf>.

[580] OECD toolkit ; WWW: <http://www.oecd-antispam.org/>.

[581] A Pisanty, 'How the Internet Organizations Were Built and Continue to Evolve'. In: WJ Drake, 'Reforming Internet Governance'; WGIG (2005). WWW: <http://www.wgig.org/docs/book/AP.html>.

[582] RC Ellickson, *Order without Law: How Neighbors Settle Disputes* (Harvard University Press, Cambridge, Mass. 1991).

important, in part for the reasons suggested here.' [583] Second argument is the vagueness of virtual creatures in a totally new media landscape.
Virtual identities, as the entity 'legal persons' generated by code, are human harebrained schemes. Many creations result in artificial fabrications that are so vague that the users of the creature are amused but the non-users want to regulate the specific phenomenon.
Arbitration could be a better way of dealing with cyber conflicts, especially concerning property rights, as government action by law enforcement may threaten property rights in cyberspace resulting in slowdowns in the growth of the cyber economy.

Wrapping up the conclusion is that new virtual and autonomous identity concepts need new legal definitions. The use of virtual identities has to be analysed against the existing regulations to find the adequate regulatory answers, in particular on basic concepts of online identity, online anonymity, on the use of multiple identities, on the regulation of identity managers and on control instruments for identity holders.

Future Expansion and Expectations

Legal View

As in history new technology is regulated by old law. Equally most elements of the Internet usage and service content were regulated in some form or fashion – prior to the arrival of the Internet. Despite the calls by some for the development of Internet-specific law, or cyberspace law – similar to the Law of the Sea – the information technology is changing so rapidly for any 'sui' generic body of law that is developing, implementing and maintaining it is a lost race against the 24/7 Internet time. Despite the technology feed, the idea is that the Internet culture is the same for all contributors, there are and will be local and cultural differentiations in compliance with the law. The 'ad hoc' way of gradual adaptations of the tried and tested fundamental legal principles, as we have seen in free speech, economics, and in privacy and intellectual property protection, is likely to be more successful. Andersen argues that mediation and arbitration can solve cyber conflicts as it does with high technology disputes. [584] A code of conduct for legal mediation can be helpful.

Kesan and Shah point out that one of the most significant theoretical advancements in the legal academy is the recognition that law is not the only method of social regulation. [585] Other methods of social control include social norms and architecture. They argue that this has led researchers in a variety of disciplines to document how the architecture of information technologies affects our online experiences and activities. The recognition of the role of architecture has led policymakers to consider architectural as well as legal solutions to societal problems. Architectural solutions utilising information technologies have been proposed for issues such as crime, competition, free speech, privacy, protection of intellectual property, and revitalising democratic discourse, they say.

Finally, ways are examined in which laws can be used to create positive ethical models in individuals and groups. It is well known by policymakers and lawmakers that the form of societal control by passing a law which restricts the undesirable behaviour is very important. If the law becomes more widely accepted, people begin to reduce misbehaviour on the principle that it is `wrong' to do so.
However, the makers of policies and laws are seldom aware of the societal structure of `cyberspace', and for this reason there is the danger that the laws they make will not create the desired ethical model, but will instead create a backlash or revolutionary movement against the society. By observing the human behaviour in virtual communities and by continuing to take time to develop realistic policies and effective laws, it is possible to avoid such a backlash.
Code looks cryptic, intangible, and isolated. But if code were transparent it could be a recognisable structure for people, such as by-laws and codes of behaviour. Increasing the openness in code will help understanding the regulation of code. Code is not exclusive for regulators or nerds. By adapting free software and increasing the modularity of the code so that what it does is evident, users can have enough control over functionality.

---

[583] BL Benson, 'International Economic Law and Commercial Arbitration', In: AN Hatzis, (ed) 'Economic Analysis of Law: a European Perspective' (2002) p 23. WWW: <http://garnet.acns.fsu.edu/~bbenson/aristid.doc>.

[584] B Andersen, 'Mediation and Arbitration of High Technology Disputes', in: AR Lodder, A Meijboom, and DTL Oosterbaan (eds) 'IT Law - The Global Future: Achievements, Plans and Ambitions'. Papers from the 20th anniversary International IFCLA conf. Amsterdam, June 1-2, 2006, (*Elsevier*, 2006).

[585] JP Kesan and RF Shah, 'Shaping code (social norms)'. *Journal of Internet Law* 9.9 (March 2006): 3(11).

Golden Code

Social control can be a guide in solving the disputes. In the popular perception, a 'gold rush' refers to a chaotic scramble for high-profit opportunities in an open-access setting, where the premium is on speed. Among law-and-economics specialists, the mining districts of the California gold rush are often cited as canonical examples of spontaneous establishment of secure property rights in the absence of legal authority. One hundred and fifty years after the Victorian internet and the Gold Rush a virtual gold search at the Cyber Internet will set the rules for data mining. ('The Goldcorp story', Tapscott & Williams) [586]

There are further similarities between the gold colonists and the cyber squatters. Both developed in their colonial period their own tailor made code, used primitive tools in the beginning, were dependent on imported propriety technologies and caused enormous changes to the region's culture and architecture. The principles "discovery" and "development" set by the Californian miners, where "discovery" is made the source of title and "development" the condition of continuance of that title, constitutes the basis of all our local laws and regulations respecting exploitation rights. Those principles also can rule for the development of the cyberworlds.

Conclusions

1) Jurisdiction

As long as the virtual world is essentially depending of real world's energy and the sock puppeteers behind the avatars (et cetera) are controlling their virtual identities from electronic connections somewhere on our planet, the virtual world will be under the (territorial) jurisdiction of the real world. The legal position of the virtual world(s) cannot be outer space, nor will it be singly local, regional, national or global sec. It will be regulated by all. This is supported by the observation that only inhabitants of our earth are using the cyberworld, that when value is involved this also finally relates to the monetary system on earth, and that the necessary network – even when it is wireless – is traceable to points of presence all over our globe. Depending on the question 'who did what where when' the jurisdiction is related to the location of each parties involved.

2) Identity Correlation

Anonymous identity in the cyberworld is – sooner or later – always legally tangible. Even an autonomic-looking virtual identity (such as the bot in the software) is always related to – at least one – human action. But – in the legal view – being the perpetrator, supervisor or owner of a computer (-system, -network) is not the same as being accountable, responsible, culpable, and/or liable. Dependent on the law in force and on the specific situation a proof of evil intent or malice aforethought is necessary to be judged as being guilty of something. The development of new investigation methods and tactics should be more multidiscipline oriented and be open for behavioural lead.

3) Identity Switch

There is not a legal but only a medial difference between an identity in the real world and that in the cyberworld. They are both human. However, it will take time and brick-by-brick jurisprudence to bridge both identity-worlds. Support on unimpeachable authority will help us: the cross-media technology leads the way to integrating both physical (journals, tools) and virtual (Internet, games) environments.

4) Autonomous VID

Virtual identities, created by a coincident of facts and autonomous executing, cannot be ruled. Normally –as with robots – they have more benefits than disadvantages. As with nature, disasters as a result of human interference (like a mud-slide after deforestation), we cannot rule everything

---

[586] D Tapscott and AD Williams, 'Innovation in the Age of Mass Collaboration'. In: *Businessweek* 01-02-2007. WWW: <http://www.businessweek.com/innovate/content/feb2007/id20070201_774736.htm>.

once it seems to be out of control. As Rustad argues: 'In contrast to a traditional crime scene, online intruders or forgers leave few digital footprints. DNA evidence, fingerprints, or other information routinely tracked in law enforcement databases are useless for investigating cybercrimes. In addition, computer records are easier to alter than paper and pencil records. Electronic robbers and forgers leave fewer clues than white-collar criminals who alter checks or intercept promissory notes.' [587] Life is fun and risky, so is virtual life too. We accept that a chess computer is an interesting opponent in order to train your mind in move and countermove, but there could be a moment that we are checkmated by the computer generated 'chess-mate'.

5) Freedom of Thought
People often express their thoughts in a virtual environment as text or images, more than they will do in real life. Cyberspace is experienced as a part of a private home. Following the discussions on the Internet concerning the attention of domestic intelligence services this kind of free thought expression could be seen as 'speech'. The evolving Web 2.0 with far-reaching individual identity management enables us to link virtual identities not only to geographic IP addresses but also to the identity of individuals. Lacking jurisprudence about *'free thought'* vs *'thought crime'* makes this a *novum*.

6) Law Making
As explained in this survey virtual worlds and illusory behaviour are of all times. The significant difference is the used medium, in combination with of the free elements of time and location. Upholding the law needs some adaptation by investigators and judges; however some hands-on experience (and using the power of the information technology) will surely help to understand the case. Many disputes in cases concerning a virtual topic can be solved by extrapolation to an equivalent in our regular physical world. Reason logically, by analogue or *a contrario* with cases in past, and learn from the past that order without law can satisfy too in particular situations.

7) Public Passkey
When the governments all over the world are creating fake 'real' and virtual identities (for instance to purge Wikipedia by forged editor identity, or to lure suspects into a trap) and are distributing false passports (to diplomats, secret agents, royal families, etcetera) the concept of a reliable Public Key Infrastructure is hard to believe and it will be even harder to convince the cheated public to expose themselves.

Summarized a virtual gold rush by virtual identities will set the exploitation codes for the data mining and transactions, as in real world. Increasingly, disputes will be solved by either arbitration, or a mix of codes, submitted by the experts, the peer groups and environmental conditions, and – if necessary – supported by official laws. Like 'open source' all code should be public and free to be empowered and improved upon by the community.

Recommendations   (With Particular Attention to the Legal Topics)

1) More Cyber Awareness
The survey shows the ways in which laws can be used to create positive ethical models in individuals and groups. It is well known by policymakers and lawmakers that the form of societal control by passing a law which restricts the undesirable behaviour is very important. If the law becomes more widely accepted, people begin to reduce misbehaviour on the principle that it is `wrong' to do so. That point of departure can be used to enter the cyberworld if the makers of policies and laws are aware of the societal structure of cyberspace. Laws they make need to create the desired ethical model in order to avoid a backlash or revolutionary movement against both the regular and virtual societies. By observing the human behaviour in virtual communities and by continuing to take time to develop realistic policies and effective laws, it is possible to avoid such a backlash.

---

[587] ML Rustad, 'Private Enforcement of Cybercrime on the Electronic Frontier,' *Southern California Interdisciplinary Law Journal*, 11(2001) p 98 in pp 63-116.

2) Artificial Intelligent Agent
The fast developing phenomenon 'artificial intelligent agents' requires sufficient regulation as this agent technology is implemented in many common transactions. Becoming part of daily life these 'outlaws' will obtain virtual identity, personality, etc.

3) Natural as Usual
In many cyber related cases courts express that they were discovering the 'unnatural' Internet. However, there is no 'nature' in the Internet. Humans build it, just as the Victorian telegraph internet. When courts understand the Internet is part of the world instead of choosing what the Internet 'should' be, they will revert to similar non-virtual cases in past.

4) Open Source Code
To understand the regulation of code, the openness in code should be increased. We should discuss who writes codes, how we can have a part in defining the code and how we might go about changing it. We should decide whether the code, that is law, should be reviewed by public. When the modularity of the code is increased so that what it does is evident, users will have enough control over functionality and bear the code. By means of virtual communities and official "blog" and "wiki" Websites (like the Internet Governance Forum) people can collectively argue and think along with the political and legal choices we have to make, such as juries in the US courts.

5) Freedom of Virtual Expression
Lawmaking (e.g. jurisprudence) is needed to prevent that virtual expression is seen as 'speech'. Otherwise virtual identities will improve in anonymity in order to prevent that people are sued by their expression of thoughts in a virtual environment.

6) Virtual Outlaws
At this moment an adrift robot is seen as a risk but not yet as a freak of nature. [588] Given the principle that laws always follow the changes in nature and society, it has to be accepted that in the virtual society also there are imbeciles and disasters. A court should consider declaring a virtual identity *non compos mentis*.

7) Cyber Jail
Instead of penalties and community services in the real world, courts could weigh the alternative of a cyber related punishment, such as excommunication from the virtual society or virtual community services as cleaning the spam surplus. Restricting people in online (trans)actions (and virtual behaviour), so called 'usage control', by automatic software ('bots') and various types of communication control, e.g. limiting the features in ability (like usage time, option, level, volume), and/or booting, blocking and banning the IP address of a connected computer.

Summarized the recommendation is that just as in the real world the expertise of the citizens in virtual worlds can be used to let the people substantiate and support the law and sentences. Lawmakers and courts also have to transform themselves to the virtual imagination.
Complex illegal looking acts by autonomous virtual identities could be handled as effects of natural imperfection, even when it is self-starting after human intervention.

Final
Before getting a MID, a Multiple Identity Disorder, all my identities wish all your virtual identities a pleasant, exiting, but also controlled stay in your personal universe...

Jacob van Kokswijk

---

[588] Guidelines running safe robotics. WWW: <http://www.fightingrobots.co.uk/documents/event_safety.pdf>.

11 Appendices

Bibliography of Related URL's
Agentlink <http://www.agentlink.org>
Analysing Legal Implications and Agent Information Systems (ALIAS) <http://www.iids.org/research/legal_aspects/>
Compulegal <http://www.compulegal.demon.nl/>
Cyber-Rights and Cyber-Liberties <http://www.cyber-rights.org/>
Foundation for Physical Agents <http://www.fipa.org>
International Association for Artificial Intelligence and Law <http://www.iaail.org/>
Institute for Knowledge and Agent Technology <http://www.cs.unimaas.nl/ikat/>
Legal Aspects of Agents Online <http://www.lea-online.net>
Software agents' Website Rik Geurts <http://www.softwareagents.nl>
UMBC Agent Web <http://agents.umbc.edu/introduction/ >

**CONTACT / INFORMATION / RESPONS
+ LATEST NEWS ABOUT THE UPDATE OF THIS BOOK BY WIKI
Visit WWW: <http://www.digital-ego.org>**

More Literature
Referred but uncited interesting publications to this topic are listed in this note: [589]

---

[589] (in chronological order)

A Wells Branscomb, *Who owns Information? From Privacy to Public Access.* (Basic Books 1994).
ME Katsh, *Law in a Digital World.* (Oxford Press 1995).
R Susskind, *Future of Law, Facing the Challenges of Information Technology.* (Oxford Press 1996).
J Rosenoer, *Cyber law, the law of the internet.* (Springer 1996).
PE Agre and R Rotenberg, (eds.) *Technology and Privacy, The New Landscape.* (MIT press 1997).
B Damer, *Avatars! Exploring and Building Virtual Worlds on the Internet.* (Peachpit Press 1997).
A Etzioni, *The limits of Privacy.* (Basic Books 1999).
M Crang, P Crang, and J May, *Virtual Geographies, Bodies Space and relations.* (Routledge 1999).
C Hine, *Virtual Ethnography.* (Sage Publ. 2000).
Y Akdeniz, C Walker, and D Wall, (eds.) *The Internet, Law and Society*, (Pearson Education 2000).
IJ Lloyd, *Legal Aspects of the Information Society.* (Butterworks 2000).
R Spinello, *Cyber Ethics, Morality and Law in Cyberspace.* (Jones & Bartlett Publishers 2000).
JJ O'Donell, *Avatars of the Word: From Papyrus to Cyberspace* (Harvard University Press; Nw Ed 2000).
JE Fountain, *Building the Virtual State, Information Technology and Institutional.* (ChangeBrookings Press 2001).
P Ludlow, (ed.) *Crypto Anarchy, Cyberstates and Pirate Utopias.* (MIT press 2001).
DS Hall, (ed.) *Crime and the internet.* (Routledge 2001).
S Biegel, Byond Our Control? Confronting the Limits of Our Legal System in the Age of Cyberspace. (MIT Press 2001).
KD Mitnick and WL Simon, *The Art of Deception: Controlling the Human Element of Security.* Wiley & Sons (2001).
J de Mul, *Cyberspace Odyssee* (Klement, Kampen 2005) (in Dutch, translated in Chinese and Turkish)
R Hunting, *Business, Crime, and Privacy in the Age of Ubiquitous Computing.* (Wiley & Sons 2002).
M Godwin, (rev. ed.) *Cyber Rights; Defending Free Speech in the Digital Age.* (MIT Press 2003).
C Nicoll, JEJ Prins, and MJM van Dellen, *Digital Anonymity and the Law tensions and Dimensions.* (Asser Press 2003).
IJ Lloyd, *Information Technology Law,* 4th ed. (Oxford press 2004).
KD Mitnick and WL Simon, *The Art of Intrusion: The Real Stories Behind the Exploits of Hackers, Intruders & Deceivers.* (Wiley & Sons 2005).
T Halbert and E Ingulli, *Cyber Ethics.* (Thomson 2005).
IJ Lloyd, *International Encyclopaedia of Laws.* (Kluwer 2006).
S Mulder, *The User Is Always Right, A practical guide to creating and using personas for the web.* (New Riders Press 1980);
S Mulder and Z Yaar, rev. ed. (New Riders Press 2006).
J Pruitt and T Adlin, *The Persona Lifecycle : Keeping People in Mind Throughout Product Design* (Morgan Kaufmann 2006).
P Kr Singh, *Laws on Cyber Crimes: Along with IT Act and Relevant Rules (*Book Enclave 2007).
R Cooper, *Alter Ego: Avatars and Their Creators* (Chris Boot, 2007).
JM Balkin et al, *Cybercrime: Digital Cops in a Networked Environment* (Ex Machina: Law, Technology, Society)(NYU Press 2007).

Email Header 1993 vs 2007

Email Header 1993
Date: Thu, 22 Jul 1993 10:54:08 -0400
Message-Id: <199307221454.AA11359@yfn.ysu.edu>
From: John D'Agostini <am500@yfn.ysu.edu>
To: subgenius@mc.lcs.mit.edu
Subject: it's raining lawyers.....
Reply-To: am500@yfn.ysu.edu

Q: What's the difference between a Toronto lawyer and a sperm ?

A: The sperm has a 1 in 100 million chance of becoming a human.

'Maude is my co-pilot'
One of the many clones of Dr. Funkenstein......

Source: WWW: <http://www.subgenius.com/subg-digest/v4/0161.html>.

Email Header 2007
From: 'VSMM 2007 Australia committe' <australia@vsmm.org>
To: <announcements@vsmm.org>
Sent: Friday, March 02, 2007 2:33 AM
Subject: [VSMM] VSMM 2007: Second call for Papers - September 23-26, 2007, Brisbane, Australia
Message: Please remind VSMM07: September 23-26, 2007, Brisbane, AU, http://australia.vsmm.org

Return-Path: <australia@vsmm.org>
Delivered-To: j.van.kokswijk@fontys.edu
Received: (qmail 19794 invoked from network); 2 Mar 2007 03:11:52 -0000
Received: from mail.redclay.us (65.39.221.118) by goliath.fontys.nl with SMTP; 2 Mar 2007 03:11:52 -0000
MIME-Version: 1.0
Date: Thu, 1 Mar 2007 17:33:26 -0800
Content-Type: text/plain; charset=iso-8859-1
Content-Transfer-Encoding: quoted-printable
Subject: [VSMM] VSMM 2007: Second call for Papers - September 23-26, 2007, Brisbane, Australia
From: VSMM 2007 Australia committe <australia@vsmm.org>
Reply-To: australia@vsmm.org
To: <announcements@vsmm.org>
CC:
Message-ID: <697476031_64694167@mail.redclay.us>
X-RBL-Warning: SUBCHARS-50: Subject with at least 50 characters found.
X-RBL-Warning: SUBCHARS-55: Subject with at least 55 characters found.
X-RBL-Warning: SUBCHARS-60: Subject with at least 60 characters found.
X-Declude-Sender: australia@vsmm.org [127.0.0.1]
X-Declude-Spoolname: 64694165.eml
X-Declude-RefID:
X-Declude-Note: Scanned by Declude 4.3.30 for spam. 'http://www.declude.com/x-note.htm'
X-Declude-Scan: Incoming Score [3] at 17:33:38 on 01 Mar 2007
X-Declude-Fail: SUBCHARS-50 [1], SUBCHARS-55 [1], SUBCHARS-60 [1]
X-Country-Chain: sender: australia@vsmm.org
X-RBL-Warning: BCC: 99 Bcc:'s detected.
X-RBL-Warning: SUBSPACE-12: Subject with at least 12 spaces found.
X-RBL-Warning: SUBCHARS-50: Subject with at least 50 characters found.
X-RBL-Warning: SUBCHARS-55: Subject with at least 55 characters found.
X-RBL-Warning: SUBCHARS-60: Subject with at least 60 characters found.
X-Declude-Sender: australia@vsmm.org [127.0.0.1]
X-Declude-Spoolname: 64694167.eml
X-Declude-RefID:
X-Declude-Note: Scanned by Declude 4.3.30 for spam. 'http://www.declude.com/x-note.htm'
X-Declude-Scan: Score [8] at 17:34:08 on 01 Mar 2007
X-Declude-Fail: BCC [4], SUBSPACE-12 [1], SUBCHARS-50 [1], SUBCHARS-55 [1], SUBCHARS-60 [1]
X-Country-Chain:

Source: private collection © 2007.